Seeds of Amazonian Plants

FERNANDO CORNEJO
AND
JOHN JANOVEC

Seeds of Amazonian Plants

PRINCETON UNIVERSITY PRESS PRINCETON AND OXFORD

Published by Princeton University Press, 41 William Street,
Princeton, New Jersey 08540

In the United Kingdom: Princeton University Press, 6 Oxford Street,
Woodstock, Oxfordshire OX20 1TW

Library of Congress Cataloging-in-Publication Data

Cornejo, Fernando, 1958–
    Seeds of Amazonian plants / Fernando Cornejo and John Janovec.
        p. cm. — (Princeton field guides)
    ISBN 978-0-691-11929-8 (hardcover : alk. paper) — ISBN 978-0-691-14647-8
(pbk. : alk. paper)   1. Tropical plants—Seeds—Amazon River Region—Identification.
    2. Plants—Amazon River Region—Identification.   3. Seeds—Amazon River Region—Morphology.
I. Janovec, John.   II. Title.
    QK241.C67 2010
    58198—dc22   2010005269

British Library Cataloging-in-Publication Data is available
This book has been composed in ITC Cheltenham and Gill Sans
Printed on acid-free paper. ∞
nathist.princeton.edu
Printed and bound in Malaysia for Imago
10   9   8   7   6   5   4   3   2   1

Fernando Cornejo dedicates this book to his family, especially his daughter Jessenia. John Janovec dedicates this book to his family, especially his wife, Madeleine, and their two boys, Silvestre and J. W. Garvie, and to the memory of Dr. Theodore M. Barkley (1936–2004).

Contents

Foreword

Tropical forests are notorious for the bewildering diversity of trees they contain. Some Amazonian and Bornean forests support more than 300 species of trees per hectare among approximately 600 trunks. Every other tree, in other words, is a species not encountered previously. At larger scales such forests often contain over 1,000 species. This formidable diversity has constituted a leading scientific mystery and at the same time has presented daunting obstacles to researchers. Mastering the diversity is a necessary first step to understanding the processes that generate and perpetuate it.

Yet that first step was slow in coming because of the formidable challenges of identifying trees in remote parts of the tropics.

As recently as the 1970s, it was conventional wisdom among botanists that the trees of highly diverse tropical forests could not be identified in the field. There were too many species and, in particular, too many that looked alike and could be confused. The only reliable way to identify trees was to collect specimens in fruit or in flower, bring them to the herbarium, and compare them, side-by-side, with specimens previously identified by specialists. This widely held opinion presented an almost insurmountable barrier to those wishing to study forests in situ. If one were to submit an article based on in-the-field identification, it would promptly be rejected by "experts" who would claim the data were unreliable because, of course, identification of trees in the field was impossible.

It took one intrepid botanist, possessed of extraordinary energy and inextinguishable optimism, to break down this barrier. Today, Alwyn Gentry, is a legend, his life tragically cut short by an airplane accident. Much of the work he did to prove to the world that tropical trees could indeed be identified in the field was carried out in the Peruvian Amazon. Now, 30 years later, field identification of tropical trees is routine and being practiced all over the world. But few people in our technologically-driven world appreciate that the newly acquired ability to identify tropical trees rests on

major intellectual advances, many of them pioneered by Gentry himself.

Before Gentry and a few others like him, botanists identified tropical trees using minute and obscure characteristics of flowers. If a tree encountered in the field wasn't flowering (more than 90 percent aren't on a given day), then forget it. You weren't going to find out what it was. Gentry changed that by studying vegetative characters in greater depth than anyone before him. Did the bark have an odor? Did it gush colored sap when nicked? Did the leaves have stipules, hairs, teeth, translucent punctations? Through careful analysis of vegetative characteristics, Gentry showed that nearly every species possesses a distinct and recognizable combination of characters that separates it from all others. These details, along with a systematic methodology for applying them, were published in a monumental book in 1993 and that book has forever changed the way researchers approach the study of tropical forests.

There is a parallel story to be told about researchers striving to understand why tropical forests are composed of so many tree species, and how the remarkable diversity of species is reconstituted generation after generation. At first, investigators began with the obvious and concentrated on the trees. But soon it became apparent that trees, being established and more or less permanent fixtures of the landscape, did not reveal many secrets. Then, beginning in 1979, the purview was extended down to the level of saplings. A new rush of insights was obtained as investigators studied the dynamics of the transition from sapling to adult, but the fundamental mystery of diversity remained intact. Now, more recently, mostly since 2000, emphasis has shifted to the earliest stages of the tree life cycle, to seeds, how seeds are distributed by dispersers, and under what circumstances seeds of each kind are most successful.

Now that research is focused on the seed-seedling-sapling transition, investigators have hit another technological barrier, for there is no way to identify seeds in the tropical forest other than by finding the tree that produced them and identifying the tree (not the seed). But seeds are scattered far and wide so the tree that produced a given seed may be hundreds of meters away. Indeed, the seeds that are carried farthest from their parents by dispersers are the ones most likely to germinate and to survive as seedlings. Thus there is a compelling need in contemporary research to be able to identify seeds independently of the trees that produced them. But identifying tropical tree seeds is not easy. The seeds come in myriad sizes, shapes and forms and there are no guides. Museum specimens preserve leaves and flowers, but not seeds. There is literally nowhere to turn for the information.

Fernando Cornejo recognized this a long time ago, in 1984, when he began a seed collection at the Cocha Cashu Biological Station in the Manu National Park, Madre de Dios, Peru. Since then the collection has grown and served as a vital reference for dozens of investigators studying forest processes at the research station. Meanwhile, Fernando has gone on to become one of Peru's leading botanists, while continuing to develop his expertise on seeds. His knowledge has expanded through a four-year stint in the tropical forest of Panama and his participation in scores of field projects and expeditions in Peru.

Fernando first met John Janovec in 1998 and in 2001 they started working together on various floristic projects in the Peruvian Amazon. No one else could have produced this volume. Fernando and John live in the Amazon and know the forest intimately. In this work, as in the book of Gentry, they are putting forward a vast knowledge accumulated over more than two decades, an unrivalled knowledge that will be eagerly received by botanists and ecologists like myself.

The text is minimal. The pictures speak for themselves for, in truth, seeds are so varied and complex in detail that words fail as a means for describing them. The key to learning how to identify seeds is similar to that to learning how to identify trees in the field. Certain distinctive characters flag particular taxonomic groups. The seeds of Sapotaceae are shiny and often flattened with a rough scar along one margin, whereas those of Anonaceae are more cylindrical and often have a sunken groove running around the middle. Characters such as these can lead one to the correct family, but from there one needs additional characters such as size, shape, color or texture to further narrow down the choices to the genus or even species level. The book guides one through these steps, though it will take some practice to know what characters are most important.

It is to be hoped that this unique volume will have an impact on future research similar to that of Gentry's volume, for it opens a window onto a new and mostly unexplored realm of forest biology.

John Terborgh
Research Professor and Director
Center for Tropical Conservation
Nicholas School, Duke University
September 21, 2009

Preface

This book is the product of more than two decades of botanical and ecological field work in the Peruvian Amazon—field work that has always included the systematic collection of seeds and other supporting plant specimens. Fernando started building a reference collection of Amazonian seeds in 1984 at the Cocha Cashu Biological Station in Manu National Park located in Madre de Dios, Peru. Subsequently he continued to build the seed collection during field work in other regions of Madre de Dios and to the north in Iquitos, Peru. We met in 1998 in Puerto Maldonado, Madre de Dios, Peru, where we briefly discussed the seed collection and Fernando's dream of producing a color field guide. In 2001 we met again and initiated what has become a long-term friendship and collaboration, with a focus on botanical inventory and ecological investigations in Madre de Dios. Our field work at various sites in the region enabled us to expand the seed collection and we began to discuss the idea of producing a field guide and what it would take to complete it. By that time we were starting to replace our 35 mm film cameras with the new point-and-shoot digital cameras, such as the Nikon Coolpix series. With a small equipment grant from the Amazon Conservation Association, and a commitment from Robert Kirk and colleagues at Princeton University Press to publish the book through their natural history field guide series, we started imaging the seed collection in July 2002. Fernando worked on finalizing the Spanish plant descriptions and I translated and improved them, mostly on paper during airplane, bus, and boat travel related to ongoing research projects in Peru.

By 2004 most of the seeds had been photographed. However, in late 2004 we learned that our initial digital images were of a lower quality and resolution than desired so we made the decision to photograph all seeds again with a newer digital camera and a better dissecting microscope. At that point we thought that everything was coming together toward the rapid production of this field guide. Little did we know, much more work would be required. We spent the next year correcting the original images to meet standards of color, contrast, size, and resolution required for final printing. Then, with a generous grant from the Gordon and Betty Moore Foundation, we initiated an intensive three-year project focused on investigations of the flora, fauna, and ecosystems at numerous sites in the Andes-Amazon region of southeastern Peru, as well as the design and programming of the Atrium Biodiversity Information System (http://atrium.andesamazon. org). As much as we wanted to finalize the book manuscript and image processing, we inevitably became focused on major field work that caused us to put this seed guide on the back burner until the end of 2007.

Fortunately, our work during 2004–2007 provided important opportunities to test the descriptions, key, and images while collecting and identifying seeds in the field. As the leading botanists working at the Los Amigos Biological Station and Conservation Area in Madre de Dios, our plant, fruit, and seed identification services were constantly requested. This gave us the opportunity to work with numerous other biologists to help them identify plants, fruits, and seeds being generated through their investigations of terrestrial and arboreal mammals such as monkeys, as well as birds. For example, in collaboration with Dr. Mathias Tobler, we used early versions of this field guide to identify hundreds of seeds that he collected from dung samples of the lowland tapir (*Tapirus terrestris*), which led to the publication of a scientific article about seeds in the diet of this species. This collaborative project and others like it enabled us to make important improvements to this field guide. We strengthened our belief that botanical data and knowledge form an essential foundation underlying all biological investigations in the complex and diverse forests of the Amazon. It became more apparent to us that our Amazon seed guide would enable our zoologist colleagues to learn to identify the seeds—and digested seed fragments—that they pull from the dung samples in the Amazon! We also hardened our dedication to botanical research; we'll forever strive to keep the ancient art and science of "doing botany" alive in this modern era when good, field-based botanists are disappearing. We still wonder, though, who will do all the work required to document the diverse flora of the Amazon and other tropical wilderness areas of the world?

The urgency of this Amazon seed guide started to drive us again during 2008–2009 as we put our minds on the task of finishing and submitting the final text and images. We overcame the constant desire to add more seed images of more plant families and genera. Our desire to make it better, to "perfect" it, to add more taxa and more seeds, changed to focusing on putting the final touches on all components of the book. By early 2009 we started submitting all final drafts of text and images to Robert Kirk and team at Princeton University Press. We then worked closely with copy editor Judith Hoffman on all text sections, and with Princeton artistic editor, Dimitri Karetnikov, on final review and editing of all images and image layout. The book was submitted to Dale Cotton, Senior Production Editor, in early October.

We are pleased to present this guide that provides a new tool for identifying the seeds of Amazonian plants. Some have asked us why we did not include more descriptive information about seeds in the plant descriptions. Based on our united experience collecting and identifying plants in Amazonian rainforests, we decided that the best approach to a seed guide is to present a whole-plant perspective with hundreds of high-quality seed images—each speaking a thousand words toward their identi-fication—with a practical identification key and whole-plant descriptions that include notes on distribution and known anthropogenic uses.

We believe this tool will be useful to a broad range of people from multiple disciplines, whose work involves seeds of the Amazon and other sites in the Neotropics. We hope that it will serve ecologists who are trying to understand the diverse plant-seed cycle and the natural regeneration of Amazonian forests. We hope that it serves horticulturists and foresters rushing to identify and propagate native plants of the Amazon. We dream that making this guide available will encourage scientists, students, and the general public to integrate their collections, data, images, interests, and knowledge of seeds and plants of the Amazon in order to enhance our understanding of the region's poorly known biological diversity. As stated in the mission of our home institution, the Botanical Research Institute of Texas (BRIT), we want this Amazon seed guide to contribute to conserving our natural heritage by deepening our understanding of the plant world and by increasing awareness about the importance that plants bring to life. We would have liked to have included more seed images and additional families and genera of Amazonian plants, but we had to cut it off at some point. We look forward to feedback as we envision future, updated versions and hope that eventually we can make this guide available in other languages, starting with Spanish and Brazilian Portuguese.

John Janovec and Fernando Cornejo
Peru, September 5, 2009

Acknowledgments

This project would not have been possible without generous support from the Botanical Research Institute of Texas (BRIT), the Gordon and Betty Moore Foundation, the U.S. National Science Foundation Biotic Surveys and Inventory Program (grant number DEB-0717453), Lucy Darden and the Discovery Fund of Fort Worth, Texas, John Mitchell and the Beneficia Foundation, Sally Channon, and the Amazon Conservation Association—thanks for believing in us and providing essential funding that enabled us to focus on completion of this Amazon seed guide. We thank Robert Kirk, Dimitri Karetnikov, and Dale Cotton of Princeton University Press, and freelancer Judith Hoffman, for their constant support, editorial and artistic guidance, and unending patience. We extend our special appreciation to Sy Sohmer, Director of BRIT, along with the Board, Administration, Development, and Staff, for institutional support and infrastructure during all phases of our work. We are indebted to the following colleagues for their artistic, technical, and editorial contributions, as well as their moral support and encouragement: Mathias Tobler, Amanda Neill, Keri McNew, Tiana Franklin, Jason Best, Renan Valega, Barney Lipscomb, Robert George, Cleve Lancaster, Pat Harrison, Bob O'Kennon, Jason Wells, Justin Allison, Frank Pachas, Marissa Oppel, Asha McElfish, Andrew Lutz, and Andrew Waltke. Keri McNew and Amanda Neill deserve extra credit for encouraging us to stay focused on completion of this book during those phases where we were juggling too many projects. We thank Marisa Ocrospoma Jara for providing the illustrations that acccompany the glossary of botanical terminology. We thank Robin Foster and John Terborgh for supporting the first author in the early stages of his botanical career and for encouraging the first seed collections toward this book. We thank Adrian Forsyth for encouraging our studies of Amazonian plants and for supporting us in early stages of the production of this book. We owe special appreciation to Piher Maceda, Angel Balarezo, Javier Huinga, and Pedro Centeno, members of the BRIT field team in Peru who assisted in various phases of plant collection activities. We thank Dr. Ernest Couch and the Department of Biology at Texas Christian University for allowing us to use their Scanning Electron Microscopy Center. We are grateful to Enrique Ortiz, Juan Carlos Flores, Nigel Pitman, and other colleagues associated with the Amazon Conservation Association (ACA) and the Asociación para la Conservación de la Cuenca Amazonica (ACCA) for encouragement and logistical support. We thank Michael Goulding for inspiration and important suggestions in early stages. We gratefully appreciate important advice and manuscript reviews from Scott Mori of the New York Botanical Garden (NYBG); his legendary knowledge of the Amazonian flora and his tireless focus and productivity has always inspired us to work diligently toward the completion of this book. We thank Nate Smith and Dennis Stevenson, also of the NYBG, for advice on plant family classification and descriptions during early phases. We appreciate the former Instituto Nacional de Recursos Naturales (INRENA) of Peru for issuing permits to carry out research and collections of the flora of the Peruvian Amazon. Last but not least, we are indebted to our families for keeping us strong and focused and for being patient with us during the production of this book.

Introduction

Though I do not believe that a plant will spring up where no seed has been, I have great faith in a seed. Convince me that you have a seed there, and I am prepared to expect wonders.
—Henry David Thoreau (*Faith in a Seed*)

The Amazonian wilderness, together with the eastern slope of the Andes, harbors the greatest terrestrial and freshwater diversity known on Earth, and is an irreplaceable resource for present and future generations. Ongoing environmental degradation, however, poses a serious threat to the region's biological diversity. This degradation is a result of numerous factors, particularly unwise exploitation of the land and rapid population growth. In megadiverse regions such as the Andes-Amazon, managing these matters and developing wise programs for conserving natural resources is a huge undertaking, requiring the efforts of an array of individuals with widely diverse expertise. An accurate accounting of the flora, combined with ecological studies, is indispensable to such efforts.

Due to their diverse forms, shapes, and sizes, seeds have always been used by collectors, artists, and naturalists. In recent years, seeds have become increasingly important to our understanding of forest regeneration, plant propagation and restoration, seed dispersal, and wildlife ecology, as well as in the identification of flora. Seeds store the genetic diversity of plants, and as the key component of the process of forest regeneration they form an essential structural component of all ecosystems. Seeds are an important food resource for animals, and have evolved diverse strategies, reflected in their shape, color, taste, odor, appendages, and protective coverings to ensure their perpetuation in an ever-changing environment. In some angiosperms the seed is part of a *diaspore* containing hairs or wings with additional plant tissue to facilitate dispersal.

During more than two centuries of natural science research in the Neotropics, biologists attempting to classify and identify the plants of this region have been met with certain challenges due to the lack of tools such as field guides or adequate identification keys, particularly guides for seeds. One seed guide has been published for Central America and northern Mexico (Lentz, 2005), and two books have been published to aid in the identification of the fruits of neotropical South America, focusing on the Guianas (Roosmalen, 1985) and the Sierra de la Macarena on the eastern slope of the central Colombian Andes (Stevenson, et al., 2000). A recent book by Lobova, et al. (2009) provides data and images for identifying seeds dispersed by bats in the Neotropics. These books exemplify the basic tools required by biologists and ecologists working in those regions. Yet to date there are no field guides for identifying the great diversity of seeds from the Amazon or other neotropical rainforests.

While this guide represents only a portion of all the genera encountered in the Amazon or the neotropical region, we have tried to include at least one example from each of the more common families that exist in the Amazonian region. Most families are well-represented by a selection of genera. For the most diverse families, such as the Orhidaceae, we were able to provide only a few representative examples of their seeds. Although the seeds of this guide were collected in the Peruvian Amazon, many of them, at the generic level, are distributed in other regions of the Neotropics, other parts of the Americas, and sometimes on other continents.

Collecting and Identifying Seeds

To properly identify seeds, the initial plant collection should be complete, preferably consisting of duplicate specimens of stems and leaves, fruits, and the seeds for deposit in reputable herbaria. If it is not possible to make stem and leaf collections when fruits and seeds are collected, this guide can be used to identify seeds with only fruit and seed characteristics. In the case of zoologists who want to identify seeds collected from animal scat, it is often impossible to collect the source plant. In such cases, especially when fruits are unknown at the time of seed collection, some generic identifications will remain tentative until further studies can be carried out in the field and herbarium. In most cases, however, based on our personal experience and field testing for this book, most seeds can be identified to the generic level by using the identification key and comparing seeds in hand to the images in this book.

When fruits are available, we recommend that images be taken in the field with notes about the fruit's size, shape, color, texture, and pubescence. Later, the seeds should be extracted from the fruits and carefully cleaned with a moist cloth. Once clean, they should be dried in an oven at no more than 50° Celsius. When dry, the seeds can be compared to the information and the images in this guide. The seeds in this book represent the dried, preserved form of the seed, similar to what one might find with an herbarium specimen or in a personal seed collection.

How to Use This Book

This book is composed of two parts, which together will help identify seeds to family or genus. The first part, AID TO THE IDENTIFICATION OF AMAZONIAN SEEDS, is a key providing basic diagnostic descriptions of the seeds and diaspores. The second part, FAMILY AND GENUS DESCRIPTIONS, includes the seed images and extends the process of identification to the whole plant. Lastly, a GLOSSARY at the end of the book contains definitions and illustrations of botanical terms used in this guide.

Aid to the Identification of Amazonian Seeds

We divide the Aid into two categories: (1) diaspores and seeds with conspicuous wings or hairs, dispersed by wind, and (2) seeds lacking wings or hairs, not dispersed by wind. Within each general category, we further divide seeds into different forms, shapes, sizes, color, and texture. Before using the Aid, it is important to understand four basic terms used to describe general seed shapes or forms:

- **Elongate:** An elongate seed is at least 1.5 times longer than it is wide. (Divide length by width; if greater than 1.5, the seed is elongate.)
- **Round:** A round seed is more uniform in length and width, appears more conspicuously circular than elongate, and length is less than 1.5 times the width. (Divide length by width; if less than 1.5, the seed is round.)
- **Flat:** A flat seed is compressed along one dimension and the width is greater than 1.4 times the thickness. (Divide width by thickness; if greater than 1.4, the seed is flat.) Thickness can also be thought of as the height along the flattened or compressed plane of the seed.
- **Irregular:** An irregular seed is one for which it is difficult or impossible to determine a specific vertical or horizontal dimension such as length, width, or thickness (height). These seeds are often variable in shape and size, even in the same fruit, and

they sometimes have a dissected, grooved, or warty texture that makes them irregular in shape and difficult to measure along specific planes.

Seeds should be measured with the ruler provided at the **end of the book** and compared with the actual dimensions presented in the image captions. Be aware that the size of fruits and seeds can vary depending on the individual fruit and the region where it is collected. At the same time, seed colors can vary depending on the state of maturity of the fruit and how the seeds have been cleaned and dried. Additionally, the forms and colors are subjective. For example, a dark brown seed could be interpreted as a black seed, or a brown seed could be interpreted as a reddish seed. If you are unable to identify your seed to genus, try a superior or inferior size range, or another seed color or form from the seed key.

Family and Genus Descriptions

In this section, we present the seed images and descriptions of Amazonian plants in alphabetical order by family according to the plant classification of Smith, et al. (2003). Although some may argue with our decision to follow a modified Cronquist (1981, 1991) classification scheme, we found this to be the most well-documented system for us to complete this book, as well for others who will be using this book to identify seeds in the Amazon and beyond.

We include brief descriptions of each plant family and genus, focusing on plant forms, leaf types and position, fruit type and position, and seed. The generic descriptions include important characteristics that were not already included in the respective family description. We also provide notes about the distribution of each genus, and for many plants we include information about some of their known uses in the Amazon.

The majority of the descriptions are based on our personal observations. However, we have also relied on some important botanical texts, especially to clarify morphological concepts and terminology for the plant descriptions, such as fruit and pubescence types, among others. The following texts served as important references in the production of this field guide:

- Dodson, C.H. and A.H. Gentry. 1978. *Flora of Rio Palenque Science Center, Los Rios Province, Ecuador.* Selby Botanical Garden Journal 4(1–6): 1–628.
- Gentry, A. H. 1993. *Field Guide to Woody Plants of Northwestern South America (Colombia, Ecuador, and Peru).* Chicago Press, Chicago, Illinois.
- Harris, J.G. and M.W. Harris. 2001. *Plant Identification Terminology: An Illustrated Glossary. Second Edition.* Spring Lake Publishing, Payson, Utah.
- Jardim, A., et al. 2003. *Guia de los Arboles y Arbustos del Bosque Seco Chiquitano Bolivia.* Missouri Botanical Garden Press, Saint Louis, Missouri.
- Lentz, D.L. and R. Dickau. 2005. *Seeds of Central America and Northern Mexico.* New York Botanical Garden Press, Bronx, NY. 310 pp.
- Lobova, T. A., et al. 2009. *Seed Dispersal by Bats in the Neotropics.* The New York Botanical Garden, Bronx, New York, USA. 471 pp. 32 color plates.
- Mori, S. A., et al. 1997. *Guide to the Vascular Plants of Central French Guiana. Part 1. Pteridophytes, Gymnosperms, and Monocotyledons.* Memoirs of the New York Botanical Garden, volume 76.
- Mori, S. A., et al. 2002. *Guide to the Vascular Plants of Central French Guiana. Part 2.*

Dicotyledons. Mem. New York Bot. Gard. 76(2): 11–776, plates 11–128.
- van Roosmalen, M. 1985. *Fruits of Guianan Flora.* University of Utrecht Press/ Veenman, Wageningen, Holland.
- Smith, N., et al. 2003. *Flowering Plants of the Neotropics.* Princeton University Press, Princeton, New Jersey.
- Stevenson, P.R., et al. 2000. *Guia de Frutos de los Bosques del Rio Duda, La Macarena, Colombia.* Asociación para la Defensa de La Macarena-IUCN (Netherlands). Santafé de Bogotá.

Seed Images

This field guide contains 750 seed images representing 543 genera in 131 families. We have attempted to include the most representative images for each genus, and in some cases we present seeds from various species, especially for the most diverse and complex genera. All seeds were dry at the time of imaging.

The images were taken indoors with a Nikon Coolpix 8500 and a Canon 20D mostly under natural light. To produce images of tiny seeds, we attached the Nikon 8500 to a standard stereomicroscope and took the images under artificial light. Some microscopic seeds were imaged with a scanning electron microscope. All original seed images included a scale bar to measure the size of the seeds. The magnification level and maximum length × width dimensions of the seeds imaged in this book are provided in the image captions. For general size reference, we include a metric scale bar on the last page of the book.

Aid to Identification of Amazonian Seeds

To enhance the seed identification process, we present the following simple but practical identification key to the seeds of plant genera included in this book. We have found it convenient to divide the complex diversity of Amazonian seeds into two main groups: (1) those dispersed by wind (Key I), and (2) those not dispersed by wind (Key II). At first this might not seem intuitive to new users, but with some preliminary information and a little practice in the field, the distinction will become obvious. Each genus in the listed in alphabetical order with its corresponding page number in the book. Sixteen genera (*Abarema, Aechmea, Aiouea, Amanoa, Centropogon, Cuphea, Dichranostyles, Hyospathe, Rhodostemnodaphne, Lacistema, Marcgravia, Graffenrieda, Maieta, Bathysa, Phoradendron,* and *Hedychium*) do not have images but are included.

I. Diaspores and Seeds with Wings or Hairs, Dispersed by Wind

Some of the most conspicuous and easy to identify seeds of the Amazon are those dispersed by wind. Wind-dispersed seeds come in two major forms, but the first step toward their identification is to determine if they have wings or hairs. If wings or hairs are present, the next step is to determine if the seed you are trying to identify is an individual seed that has fallen from a distant fruit, or if it is actually a *diaspore*—a structure that fuses one or more seeds and the surrounding fruiting body with winged or hairy appendages, forming one dispersal unit.

A diaspore usually has an obvious fruiting stalk (peduncle) connected to the appendages that surround, support, or engulf the seed—this might be the easiest way for a novice to identify a true diaspore in hand. The combined seed and fruit structure breaks at the stalk from the mother plant to be picked up by the wind. Sometimes the seed or seeds can be easily removed or broken away from the winged or hairy appendages, or from the peduncle connected to those appendages. Sometimes multiple seeds are clustered together and supported from beneath by modified, wing-like leaves (bracts), and dispersed as one unit.

On the contrary, a non-diaspore winged or hairy seed is not attached to a fruiting body or stalk—the seed falls from a fruiting body that opens at maturity to release the wind-dispersed seed. The wings or hairs are arranged in one form or another on or around the individual seed.

IA. Diaspores with wings or hairs

- **Diaspores with a circular wing beneath or surrounding the seed:** *Calycobolus* (45), *Chaunochiton* (106), *Hiraea* (87), *Mascagnia* (87), and *Pterocarpus* (68).
- **Diaspores with wings on both sides of the seed:** *Cedrelinga* (60), *Dalbergia* (62), *Diplotropis* (63), *Entadopsis* (64), *Gouania* (115), *Phyllocarpus* (68), *Sclerolobium* (69), *Tachigali* (70), *Terminalia* (43), and *Toulicia* (129).
- **Diaspores with a wing on one side of the seed:** *Banisteriopsis* (86), *Gallesia* (109), *Heteropterys* (87), *Machaerium* (66), *Myroxylon* (67), *Securidaca* (113), *Seguieria* (110), *Serjania* (128), *Thinouia* (129), and *Vatairea* (70).
- **Diaspores with 3–5 wings:** *Astronium* (3), *Cardiospermum* (127), *Cavanillesia* (27), *Combretum* (43), *Cordia* (30), *Dicella* (87), *Erisma* (146), *Merremia* (46), *Petrea* (144), *Ruprechtia* (114), *Triplaris* (114), *Terminalia* (43), and *Tetrapterys* (88).
- **Diaspores with hairs:** *Guazuma* (138) and *Heliocarpus* (140).

IB. Seeds with Wings

- **Seeds with a circular wing:** *Aristolochia* (21), *Aspidosperma* (9), *Himatanthus* (10), *Jacaranda* (25), *Physocalymma* (85), *Pithecoctenium* (26), and *Manettia* (121).
- **Wings on both sides of the seed:** *Anemopaegma* (24), *Arrabidaea* (24), *Calycophyllum* (117), *Capirona* (117), *Centrolobium* (60), *Clytostoma* (25), *Couratari* (82), *Dist-*

ictella (25), *Lafoensia* (85), *Ladenbergia* (120), *Marila* (41), *Mussatia* (26), *Paragonia* (26), *Perriarrabidaea* (26), *Pseudopiptadenia* (68), *Pleonotoma* (26), *Roupala* (114), *Siolmatra* (48), *Tabebuia* (27), *Tanaecium* (27), *Uncaria* (124), and *Warscewiczia* (124).

• **Wing on one side of the seed:** *Amburana* (59), *Anthodon* (50), *Caraipa* (14), *Cariniana* (56), *Cedrela* (67), *Hippocratea* (50), *Huberodendron* (28), *Lueheopsis* (115), *Mollia* (115), *Pristimera* (51), *Pterygota* (138), *Qualea* (120), *Schizolobium* (43), *Swietenia* (94), *Simira* (98), and *Vochysia* (121).

IC. Seeds with Hairs

• **Hairs on one side of the seed:** *Asteraceae* (21), *Forsteronia* (10), *Guzmania* (32), *Hillia* (120), *Marsdenia* (21), *Odontadenia* (11), *Prestonia* (12), and *Tillandsia* (32).

• **Hairs** surrounding **the margin of the seed:** *Cochlospermum* (42) and *Ipomoea* (46).

Hairs thick, covering the seed in the form of cotton: *Ceiba* (28), *Chorisia* (28), *Eriotheca* (28), *Ochroma* (29), *Pfaffia* (2), Pseudobombax (30), and *Salix* (126).

II. Seeds Lacking Wings or Hairs, Not Dispersed by Wind

IIA. Seeds Tiny (< 0.5 cm)

IIAa. Seeds Flat, Gray or Blackish

• **Seeds with an irregular or striate surface:** *Piper* (110).
• **Seeds with a smooth, opaque surface:** *Commelina* (44) and *Senna* (69).
• **Seeds with a smooth, glossy surface:** *Chamissoa* (2), *Pera* (57), and *Phytolacca* (109).

IIAb. Seeds Flat, Brown or Reddish

• **Seeds with an irregular or striate surface:** *Brunfelsia* (134), *Centropogon* (34), *Cuphea* (85), *Geophila* (119), *Gonzalagunia* (119), *Juncus* (80), *Lantana* (143),

Mayaca (90), *Palicourea* (121), *Phoradendron* (145), *Psychotria* (122), *Siparuna* (133), *Solanum* (136), *Stenostephanus* (1), *Piper* (110), and *Urera* (142).
• **Seeds with a pubescent surface:** *Axonopus* (111).
• **Seeds with a smooth, opaque surface:** *Capsicum* (134), *Carludovica* (48), *Cuphea* (85), *Lycianthes* (135), *Mimosa* (66), *Panicum* (112), *Paspalum* (112), *Physalis* (136), *Psidium* (104), *Ruellia* (1), *Solanum* (136), *Thoracocarpus* (48), and *Tournefortia* (31).
• **Seeds with a smooth, glossy surface:** *Calopogonium* (60), *Crotalaria* (61), and *Gouania* (115).

IIAc. Seeds Flat, Whitish or Yellowish

• **Seeds with a smooth, opaque surface:** *Lycianthes* (135) and *Margaritaria* (56).

IIAd. Seeds Irregular, Gray or Blackish

• **Seeds with an irregular or striate surface:** *Bertiera* (117), *Epiphyllum* (34), *Floscopa* (44), *Maprounea* (56), *Piper* (110), and *Salpinga* (92).

IIAe. Seeds Irregular, Brown or Reddish

• **Seeds with an irregular or striate surface:** *Bathysa* (117), *Amaioua* (116), *Cuphea* (85), *Guazuma* (138), *Hamelia* (119), *Hylaeanthe* (89), *Ludwigia* (107), *Maieta* (92), *Miconia* (92), *Oldenlandia* (121), *Pentagonia* (122), *Phoradendron* (145), *Sabicea* (124), and *Xiphidium* (75).
• **Seeds with a pubescent surface:** *Ryania* (73).
• **Seeds with a smooth, opaque surface:** *Apeiba* (140), *Besleria* (74), *Billbergia* (31), *Cuphea* (85), *Ficus* (98), *Henriettella* (91), *Hibiscus* (88), *Ombrophytum* (24), *Prockia* (73), and *Xylosma* (73).
• **Seeds with a smooth, glossy surface:** *Aeschynomene* (58), *Dimerocostus* (47), *Clitoria* (61), and *Costus* (46).

IIAf. Seeds Irregular, Whitish or Yellowish

• **Seeds with a smooth, opaque surface:** *Ficus* (98), *Graffenrieda* (91), and *Miconia* (92).

IIAg. Seeds Elongate or Elliptic, Gray or Blackish

- **Seeds with an irregular or striate surface:** *Ananas* (31), *Hedyosmum* (38), and *Selenicereus* (34).
- **Seeds with a smooth, opaque surface:** *Vismia* (42).

IIAh. Seeds Elongate or Elliptic, Brown or Reddish

Seeds with an irregular or striate surface: *Anthurium* (13), *Banara* (71), *Begonia* (24), *Bellucia* (91), *Cecropia* (37), *Coussapoa* (37), *Eryngium* (9), *Fuschia* (106), *Hawkesiophyton* (135), *Hedyosmum* (38), *Norantea* (90), *Sagittaria* (2), *Scoparia* (132), and *Xanthosoma* (15).
- **Seeds with a smooth, opaque surface:** *Drymonia* (74), *Havetiopsis* (40), *Hyptis* (80), *Juanulloa* (135), *Loreya* (91), *Ludwigia* (107), *Philodendron* (14), *Potalia* (83), *Piper* (110), *Sida* (88), *Souroubea* (90), *Vismia* (42), and *Xyris* (147).

IIAi. Seeds Elongate or Elliptic, Whitish or Yellowish

- **Seeds with an irregular or striate surface:** *Anthurium* (13) and *Tibouchina* (93).
- **Seeds with a smooth, opaque surface:** *Adelobotrys* (90), *Dendropanax* (15), and *Wulffia* (23).

IIAj. Seeds Round, Gray or Blackish

- **Seeds with an irregular or striate surface:** *Alchornea* (53).
- **Seeds with a smooth, glossy surface:** *Polygonum* (114), *Scleria* (49), *Tetracera* (50), and *Zanthoxylum* (125).

IIAk. Seeds Round, Brown or Reddish

- **Seeds with an irregular or striate surface:** *Acalypha* (52), *Ardisia* (102), *Casearia* (71), *Cleome* (35), *Cuphea* (85), *Eryngium* (9), *Hyeronima* (55), *Lunania* (72), *Ossaea* (92), *Piper* (110), *Trema* (142), and *Urera* (142).
- **Seeds with a smooth, opaque surface:**

Cissus (145), *Clibadium* (22), *Croton* (53), *Cuphea* (85), *Laetia* (72), *Philodendron* (14), *Piper* (110), and *Vigna* (70).
- **Seeds with a smooth, glossy surface:** *Lasiacis* (111).

IIAl. Seeds Round, Whitish or Yellowish

- **Seeds with an irregular or striate surface:** *Dorstenia* (98).
- **Seeds with a smooth, glossy surface:** *Scleria* (49).

IIB. Small Seeds (0.5–0.99 cm)

IIBa. Seeds Flat, Gray or Black

- **Seeds with an irregular or striate surface:** *Doliocarpus* (50), *Passiflora* (109), *Sapium* (57), and *Siparuna* (133).
- **Seeds with a smooth, opaque surface:** *Acacia* (58) and *Hasseltia* (72).
- **Seeds with a smooth, glossy surface:** *Centrosema* (61), *Pera* (57), and *Trichostigma* (110).

IIBb. Seeds Flat, Brown

- **Seeds with an irregular or striate surface:** *Cestrum* (134), *Geophila* (119), *Morinda* (121), *Palicourea* (121), *Passiflora* (109), and *Psychotria* (122).
- **Seeds with a pubescent surface:** *Capparis* (34), *Desmodium* (62), and *Polygala* (113).
- **Seeds with a smooth, opaque surface:** *Alibertia* (116), *Apuleia* (59), *Borojoa* (117), *Calliandra* (59), *Calycophysum* (47), *Cyphomandra* (134), *Duroia* (118), *Gavarretia* (54), *Genipa* (119), *Gurania* (47), *Inga* (65), *Ixora* (120), *Palicourea* (121), *Psychotria* (122), *Rudgea* (123), and *Senna* (69).
- **Seeds with a smooth, glossy surface:** *Dialium* (62) and *Dracontium* (13).

IIBc. Seeds Flat, Whitish or Yellowish

- **Seeds with an irregular or striate surface:** *Cestrum* (134), *Cissus* (145), and *Mayna* (73).
- **Seeds with a pubescent surface:** *Melothria* (48).
- **Seeds with a smooth, opaque surface:**

Cyphomandra (134), *Psidium* (104), *Psychotria* (122), and *Schefflera* (16).
• **Seeds with a smooth, glossy surface:** *Turpinia* (137) and *Macrosamanea* (66).

IIBd. Seeds Irregular, Gray or Blackish

• **Seeds with an irregular or striate surface:** *Dichorisandra* (44).
• **Seeds with a smooth, opaque surface:** *Coccoloba* (113) and *Posoqueria* (122).

IIBe. Seeds Irregular, Brown or Reddish

• **Seeds with an irregular or striate surface:** *Marcgravia* (89).
• **Seeds with a smooth, opaque surface:** *Alibertia* (116), *Bixa* (27), *Dieffenbachia* (13), *Lindackeria* (72), and *Marcgravia* (89).
• **Seeds with a smooth, glossy surface:** *Renealmia* (147).

IIBf. Seeds Irregular, Whitish or Yellowish

• **Seeds with an irregular or striate surface:** *Carica* (36), *Carpotroche* (71), and *Casearia* (71).

IIBg. Seeds Elongate or Elliptic, Gray or Blackish

• **Seeds with an irregular or striate surface:** *Cassiporea* (116) and *Neoregelia* (32).
• **Seeds with a smooth, opaque surface:** *Chrysochlamys* (40), *Commelina* (44), *Erythrina* (64), *Monstera* (14), and *Syngonium* (15).

IIBh. Seeds Elongate or Elliptic, Brown or Reddish

• **Seeds with an irregular or striate surface:** *Anthurium* (13), *Ferdinandusa* (119), *Jacaratia* (36), *Marcgravia* (89), *Picramnia* (132), *Rollinia* (8), and *Trichilia* (94).
• **Seeds with a smooth, opaque surface:** *Aegiphila* (143), *Hedychium* (147), *Lacmellea* (11), *Markea* (135), *Neea* (105), *Pavonia* (88), *Pleurostachys* (49), *Protium* (32), *Rinorea* (145), *Styloceras* (33), and *Rinoreocarpus* (145).

• **Seeds with a smooth, glossy surface:** *Diplasia* (49) and *Myrcia* (104).

IIBi. Seeds Elongate or Elliptic, Whitish or Yellowish

• **Seeds with an irregular or striate surface:** *Psychotria* (122).
• **Seeds with a smooth, opaque surface:** *Abarema* (58), *Annona* (5), *Citharexylum* (143), *Faramea* (118), *Hedychium* (147), *Hyospathe* (18), and *Pseudolmedia* (100).
• **Seeds with a smooth, glossy surface:** *Acroceras* (111) and *Olyra* (112).

IIBj. Seeds Elongate or Elliptic, Blueish

• **Seeds with smooth, glossy surface:** *Abarema* (58).

IIBk. Seeds Round, Gray or Blackish

• **Seeds with an irregular or striate surface:** *Poraqueiba* (79).
• **Seeds with a pubescent surface:** *Bactris* (17).
• **Seeds with a smooth, opaque surface:** *Gymnosporia* (38) and *Lepidocaryum* (19).
• **Seeds with a smooth, glossy surface:** *Averrhoideum* (127), *Canna* (34), *Matayba* (127), and *Rhynchosia* (69).

IIBl. Seeds Round, Brown or Reddish

• **Seeds with an irregular or striate surface:** *Byrsonima* (86), *Cremastosperma* (5), *Hebepetalum* (77), *Stylogyne* (102), and *Trattinnickia* (33).
• **Seeds with a pubescent surface:** *Rinorea* (145).
• **Seeds with a smooth, opaque surface:** *Adenophaedra* (53), *Geonoma* (18), *Hedychium* (147), *Myrciaria* (104), *Pausandra* (56), *Perebea* (100), *Smilax* (133), and *Trichilia* (94).
• **Seeds with a smooth, glossy surface:** *Calyptranthes* (103), *Diplasia* (49), and *Myrcia* (104).

IIBm. Seeds Round, Whitish or Yellowish

• **Seeds with an irregular or striate surface:** *Cissampelos* (96), *Cremastosperma* (5), and *Strychnos* (84).

• **Seeds with a smooth, opaque surface:** *Cybianthus* (102), *Faramea* (118), *Gloeospermum* (144), *Hedychium* (147), *Magnolia* (85), and *Perebea* (100).

IIC. Medium Seeds (1–1.99 cm)

IICa. Seeds Flat, Gray or Blackish

• **Seeds with an irregular or striate surface:** *Cissus* (145) and *Mayna* (73).
• **Seeds with a pubescent surface:** *Couroupita* (83).
• **Seeds with a smooth, opaque surface:** *Acacia* (58) and *Cayaponia* (47).

IICb. Seeds Flat, Brown

• **Seeds with an irregular or striate surface:** *Campomanesia* (103), *Crataeva* (35), *Dioscorea* (51), *Kotchubaea* (120), *Maytenus* (38), *Petiveria* (109), *Rauvolfia* (12), *Strychnos* (84), and *Unonopsis* (9).
• **Seeds with a pubescent surface:** *Disciphania* (96).
• **Seeds with a smooth, opaque surface:** *Bauhinia* (59), *Byttneria* (137), *Caesalpinia* (59), *Celtis* (141), *Dioclea* (62), *Hura* (55), *Inga* (65), *Parkia* (68), *Randia* (123), and *Zygia* (71).
• **Seeds with a smooth, glossy surface:** *Manilkara* (130).

IICc. Seeds Flat, Whitish or Yellowish

• **Seeds with an irregular or striate surface:** *Discophora* (79) and *Morisonia* (35).
• **Seeds with a pubescent surface:** *Capparis* (34) and *Helicostylis* (99).
• **Seeds with a smooth, opaque surface:** *Averrhoa* (108), *Botryarrhena* (117), *Geissospermum* (10), *Guadua* (111), *Inga* (65), and *Ixora* (120).
• **Seeds with a smooth, glossy surface:** *Enterolobium* (64).

IICd. Seeds Irregular, Gray or Blackish

• **Seeds with an irregular or striate surface:** *Heliconia* (75) and *Dicranostyles* (46).

• **Seeds with a smooth, glossy surface:** *Anaxagorea* (4) and *Dicranostyles* (46).

IICe. Seeds Irregular, Brown

• **Seeds with an irregular or striate surface:** *Guarea* (94) and *Dicranostyles* (46).
• **Seeds with a smooth, opaque surface:** *Aiouea* (80), *Amanoa* (53), *Dicranostyles* (46), *Swartzia* (70), and *Thiloa* (43).
• **Seeds with a smooth, glossy surface:** *Swartzia* (70),

IICf. Seeds Irregular, Whitish or Yellowish

• **Seeds with an irregular or striate surface:** *Herrania* (138) and *Tabernaemontana* (12).
• **Seeds with a smooth, opaque surface:** *Batocarpus* (97), *Eugenia* (103), and *Mouriri* (95).

IICg. Seeds Elongate or Elliptic, Gray or Black

• **Seeds with an irregular or striate surface:** *Dendrobangia* (79), *Dicranostyles* (46), *Heliconia* (75), *Heteropsis* (14), *Macoubea* (11), and *Parahancornia* (12).
• **Seeds with a pubescent surface:** *Desmoncus* (18) and *Vitex* (144).
• **Seeds with a smooth, opaque surface:** *Dicranostyles* (46), *Monotagma* (89), *Monstera* (14), and *Xylopia* (9).
• **Seeds with a smooth, glossy surface:** *Connarus* (45), *Eucharis* (2), *Paullinia* (127), *Pseudoconnarus* (45), and *Rourea* (45).

IICh. Seeds Elongate or Elliptic, Brown

• **Seeds with an irregular or striate surface:** *Aiouea* (80), *Ampelocera* (141), *Dicranostyles* (46), *Dilkea* (108), *Drypetes* (54), *Guarea* (94), *Heisteria* (106), *Hirtella* (39), *Ouratea* (105), *Oxandra* (7), *Simarouba* (132), *Roucheria* (77), *Rourea* (45), and *Tapirira* (4).
• **Seeds with a pubescent surface:** *Bactris* (17), *Pharus* (112), *Quararibea* (30), and *Quiina* (115).
• **Seeds with a smooth, opaque surface:** *Aiouea* (80), *Annona* (5), *Buchenavia* (42),

Chamaedorea (17), Cordia (30), Curarea (96), Dialypetalanthus (50), Dicranostyles (46), Diospyros (51), Erythroxylum (52), Esenbeckia (125), Eugenia (103), Fusaea (6), Guatteria (6), Leonia (144), Mendoncia (1), Mollinedia (97), Moutabea (113), Myrcia (104), Nectandra (81), Neea (105), Ocotea (81), Paullinia (127), Pourouma (37), Protium (32), Sarcaulus (131), Sloanea (51), Talisia (128), Tapura (50), Tetragastris (33), Tovomita (41), and Trigynaea (8).
• **Seeds with a smooth, glossy surface:** Cymbopetalum (5) and Pouteria (130).

IICi. Seeds Elongate or Elliptic, Red or Reddish

• **Seeds with an irregular or striate surface:** Guatteria (6) and Malmea (7).
• **Seeds with a smooth, opaque surface:** Erythrina (64).

IICj. Seeds Elongate or Elliptic, Whitish or Yellowish

• **Seeds with an irregular or striate surface:** Chomelia (118), Coussarea (118), Erythroxylum (52), Ilex (12), Iriartella (19), Klarobelia (7), Malanea (7), Odontocarya (96), and Pholidostachys (20).
• **Seeds with a pubescent surface:** Chamaedorea (17), Quararibea (30), and Rollinia (8).
• **Seeds with a smooth, opaque surface:** Aegiphila (143), Borismene (95), Magnolia (85), Mouriri (95), Mucoa (11), Naucleopsis (100), Pourouma (37), Sparattanthelium (75), and Vitex (144).
• **Seeds with a smooth, glossy surface:** Richeria (57).

IICk. Seeds Round, Gray or Blackish

• **Seeds with an irregular or striate surface:** Heteropsis (14).
• **Seeds with a pubescent surface:** Oenocarpus (20).
• **Seeds with a smooth, opaque surface:** Copaifera (61) and Phenakospermum (139).
• **Seeds with a smooth, glossy surface:** Ormosia (67), Paullinia (127), and Sapindus (128).

IICl. Seeds Round, Brown

• **Seeds with an irregular or striate surface:** Allophylus (126), Mauritiella (19), Pseudoxandra (8), Sciadotenia (96), Sloanea (51), and Virola (102).
• **Seeds with a smooth, opaque surface:** Amanoa (53), Brosimum (97), Calyptranthes (103), Cordia (30), Geonoma (18), Mabea (55), Manihot (55), Mendoncia (1), Mollinedia (97), Nealchornea (56), Nectandra (81), Ocotea (81), Protium (32), Senefeldera (58), Trimatococcus (101), and Zizyphus (115).
• **Seeds with a smooth, glossy surface:** Amanoa (53), Conceveiba (53), Cupania (127), Cymbopetalum (5), Ricinus (57), and Ruizodendron (8).

IICm. Seeds Round, Red or Reddish

• **Seeds with an irregular or striate surface:** Oxandra (7).
• **Seeds with a smooth, opaque surface:** Cabralea (93) and Clavija (139).
• **Seeds with a smooth, glossy surface:** Ormosia (67).

IICn. Seeds Round, Whitish, or Yellowish

• **Seeds with an irregular or striate surface:** Celtis (141), Coussarea (118), Iriartella (19), and Schoenobiblus (140).
• **Seeds with a pubescent surface:** Euterpe (18).
• **Seeds with a smooth, opaque surface:** Agonandra (107), Borismene (95), Brosimum (97), Castilla (98), Chelyocarpus (17), Huertea (137), Maquira (99), Maripa (46), Meliosma (125), Prunus (116), Sorocea (100), and Talauma (85).

IID. Large Seeds (> 2 cm)

IIDa. Seeds Flat, Gray or Blackish

• **Seeds with an irregular or striate surface:** Antrocaryon (3).
• **Seeds with a pubescent surface:** Spondias (3).
• **Seeds with a smooth, opaque surface:** Anthodiscus (36), Canavalia (60), and Parkia (68).

- **Seeds with a smooth, glossy surface:** *Mucuna* (66).

IIDb. Seeds Flat, Brown

- **Seeds with an irregular or striate surface:** *Onychopetalum* (7) and *Strychnos* (84).
- **Seeds with a smooth, opaque surface:** *Crudia* (61), *Dioclea* (62), and *Porcelia* (7).
- **Seeds with a smooth, glossy surface:** *Chrysophyllum* (129) and *Mucuna* (66).

IIDc. Seeds Flat, Whitish or Yellowish

- **Seeds with an irregular or striate surface:** *Fevillea* (47) and *Strychnos* (84).
- **Seeds with a smooth, opaque surface:** *Bunchosia* (86).

IIDd. Seeds Irregular, Brown

- **Seeds with an irregular or striate surface:** *Bertholletia* (82).
- **Seeds with a smooth, opaque surface:** *Anacardium* (3), *Eschweilera* (83), *Gustavia* (83), *Rhodostemonodaphne* (82), and *Zygia* (71).

IIDe. Seeds Irregular, Whitish or Yellowish

- **Seeds with a pubescent surface:** *Pacouria* (11).
- **Seeds with a smooth, opaque surface:** *Phytelephas* (20).

IIDf. Seeds Elongate or Elliptic, Gray or Blackish

- **Seeds with an irregular or striate surface:** *Calatola* (78), *Caryocar* (36), and *Salacia* (77).
- **Seeds with a pubescent surface:** *Couepia* (39) and *Oenocarpus* (20).
- **Seeds with a smooth, opaque surface:** *Clarisia* (98), *Psittacanthus* (84), *Rhodostemonodaphne* (82), *Sterculia* (139), and *Tovomita* (41).
- **Seeds with a smooth, glossy surface:** *Diclinanona* (6).

IIDg. Seeds Elongate or Elliptic, Brown

- **Seeds with an irregular or striate surface:** *Abuta* (95), *Drypetes* (54), *Iryan-*

thera (101), *Mauritia* (19), *Osteophloeum* (101), *Sacoglottis* (78), *Symphonia* (41), *Talisia* (128), and *Vantanea* (78).
- **Seeds with a pubescent surface:** *Bactris* (17), *Dipteryx* (63), and *Matisia* (29).
- **Seeds with a smooth, opaque surface:** *Aniba* (80), *Beilschmiedia* (81), *Cheiloclinium* (76), *Chrysophyllum* (129), *Duguetia* (6), *Dussia* (63), *Endlicheria* (81), *Garcinia* (40), *Hymenaea* (65), *Lecointea* (66), *Pachira* (29), *Peritassa* (76), *Rheedia* (41), *Rhodostemonodaphne* (82), *Salacia* (77), *Simaba* (132), *Sloanea* (51), *Theobroma* (139), *Thiloa* (43), *Tontelea* (77), *Ischnosiphon* (89), and *Wettenia* (21).
- **Seeds with a smooth, glossy surface:** *Enterolobium* (64) and *Pouteria* (130).

IIDh. Seeds Elongate or Elliptic, Whitish or Yellowish

- **Seeds with an irregular or striate surface:** *Anomospermum* (95), *Glycidendron* (54), *Grias* (83), *Minquartia* (106), and *Spondias* (3).
- **Seeds with a pubescent surface:** *Attalea* (16) and *Matisia* (29).
- **Seeds with a smooth, opaque surface:** *Dacryodes* (32), *Garcinia* (40), *Gnetum* (74), and *Maquira* (99).

IIDi. Seeds Round, Gray or Blackish

- **Seeds with an irregular or striate surface:** *Calatola* (78).
- **Seeds with a smooth, opaque surface:** *Licania* (39) and *Omphalea* (56).

IIDj. Seeds Round, Brown

- **Seeds with an irregular or striate surface:** *Drypetes* (54), *Juglans* (79), *Mauritia* (19), *Osteophloeum* (101), *Otoba* (101), *Sloanea* (51), *Socratea* (20), and *Symphonia* (41).
- **Seeds with a pubescent surface:** *Caryocar* (36).
- **Seeds with a smooth, opaque surface:** *Calophyllum* (39), *Croton* (53), *Eugenia* (103), *Hevea* (54), *Hymenaea* (65), *Licania* (39), *Ophiocaryon* (126), and *Pachira* (29).

IIDk. Seeds Round, Whitish or Yellowish

• **Seeds with a irregular or striate surface:** *Anomospermum* (95), *Casimirella*
(78), *Cryptocarya* (81), *Iriartea* (19), and *Spondias* (3).

• **Seeds with a pubescent surface:** *Astrocaryum* (16) and *Spondias* (3).

Family and Genus Descriptions

ACANTHACEAE

Herbs, shrubs, and lianas, rarely small trees. Stems often with swollen nodes, most conspicuous when dry. Leaves simple, mostly opposite, rarely alternate. Stems and leaves often with tiny cystoliths ("stone cells") appearing conspicuously as white punctations. Fruit a bivalved capsule with longitudinal or loculicidal dehiscence, or a drupe. Some genera recognizable by showy inflorescences with large bracts or bracteoles of different colors.

Mendoncia Vell. ex Vand. Vines and lianas. Leaves simple, opposite, entire. Cystoliths not apparent. Infructescences axillary. Fruit a drupe to 2.5 cm long, black when mature, surrounded by glabrous or pubescent calyx lobes; the mesocarp fleshy, dark purple to blackish when mature. Seeds one per fruit. Some authors separate this genus into the family Mendonciaceae. Distribution: Central America to Bolivia and in the Old World tropics.

Acanthaceae - *Mendoncia* (2.31x; 12.39 × 7.34)

Ruellia L. Herbs to 2 m tall. Leaves simple, opposite, entire. Stems and leaves with cystoliths. Infructescence axillary. Fruit a bivalved capsule, to 2 cm long, brown at maturity. Seeds numerous per fruit, remaining attached to the funiculus. Distribution: Central America to Paraguay and Argentina.

Acanthaceae - *Mendoncia* (2.39x; 16.77 × 8.05)

Acanthaceae - *Ruellia* (6.88x; 4.42 × 4.15)

Stenostephanus Nees. Shrubs to 3 m tall. Leaves simple, opposite, entire; petioles short; laminas large. Infructescence terminal. Fruit a bivalved capsule, to 2 cm long, brown when mature. Seeds numerous per fruit, remaining attached to the funiculus. Distribution: Peru.

Acanthaceae - *Mendoncia* (2.73x; 13.53 × 4.96)

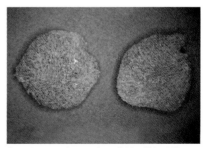

Acanthaceae - *Stenostephanus* (6.19x; 4.56 × 4.12)

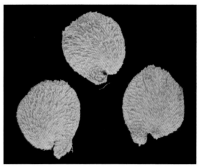

Acanthaceae - *Stenostephanus* (5.82×; 4.6 × 3.8)

ALISMATACEAE

Aquatic or subaquatic herbs. Latex present in some species. Leaves alternate, linear, floating or submerged in some species; venation parallel. Infructescence axillary or terminal. Fruit an achene.

Sagittaria L. Aquatic herbs. Leaves basal, linear, submerged in some species; venation parallel. Infructescence terminal. Fruit a dry capsule, brown when mature. Seeds tiny, numerous per fruit. Roots consumed as a starch product in some regions of the Americas. Distribution: Worldwide, typically growing in wetland habitats.

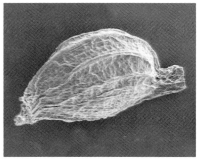

Alismataceae - *Sagittaria* (136.51×; 0.43 × 0.22)

AMARANTHACEAE

Primarily lianas and prostrate herbs, some subshrubs. Leaves simple, mostly opposite, entire; only two genera in our region with alternate leaves. Infructescence with numerous tiny bracts. Fruit an utricle.

Chamissoa Kunth. Lianas. Leaves alternate, entire. Infructescence terminal and axillary. Fruit an utricle to 0.5 cm diameter, red when mature, the valves falling when mature. Seeds one per fruit, surrounded by a white aril. Distribution: Central Mexico to northern Argentina.

Amaranthaceae - *Chamissoa* (12.01×; 2.15 × 2.04)

Pfaffia Mart. Lianas. Leaves opposite, entire. Infructescence terminal and axillary. Fruit an utricle to 0.8 cm diameter, golden when mature, the valves falling when mature covering of the seed. Seeds one per fruit, surrounded by white woolly pubescence. Similar to the genus *Iresine*, differing technically only by the stamens. Distribution: Mexico to Peru.

Amaranthaceae - *Pfaffia* (9.68×; 2.12 × 1.45)

AMARYLLIDACEAE

Herbs, mostly terrestrial, sometimes aquatic. Stems absent. Leaves emerging from a subterranean bulb. Leaves simple, succulent, sheathing the stem; venation parallel. Fruit a loculicidal capsule. Many species are cultivated as ornamental plants for their showy flowers.

Eucharis Planch. & Linden. Herbs to 0.8 m tall. Leaves linear, very thin, nearly transparent when dry; venation parallel, conspicuous. Infructescence emerging from a subterranean bulb, developing on a

Amaryllidaceae - *Eucharis* (2.48×; 11.84 × 7.01)

long peduncle. Fruit a trivalved capsule, to 1.5 cm diameter, orange at maturity. Seeds 1–5 per fruit, the endosperm metallic black and fragile. Distribution: Brazil and Peru.

ANACARDIACEAE

Shrubs and trees. Resin aromatic and conspicuous, especially from trunk and fruit. Leaves imparipinnately compound, rarely simple in a few genera, always alternate. Infructescence paniculate, axillary, or terminal. Fruit typically a drupe. Vegetatively easy to confuse with other families that have compound leaves, especially Meliaceae and Burseraceae. Many genera with edible fruits.

Anacardium L. Shrubs and trees to 30 m tall. Leaves simple. Infructescence terminal. Fruit a drupe, kidney-shaped, to 3.5 cm long, brown when mature, exposed at the end of the peduncle; the peduncle large, swollen, carnose, and aromatic, yellow or red when mature. Seeds one per fruit. One species (*A. occidentale*) is cultivated in many regions of the world as the commercial source of the cashew nut. The swollen peduncle, also referred to as a hypocarp, is edible and used to make cashew wine in some countries. Distribution: Native to the American tropics.

Anacardiaceae - *Anacardium* (1.9×; 26.33 × 21)

Anacardiaceae - *Anacardium* (1.88×; 29.1 × 20.34)

Antrocaryon Pierre. Trees to 35 m tall. Leaves compound, the leaflets thick, fragile when dry. Fruit a fleshy drupe to 2 cm diameter, yellow when mature, edible with rich flavor. Seeds one per fruit. A little known genus related to *Spondias*, with similar trunk and fruit. Distribution: Brazil, Colombia, and Peru.

Anacardiaceae - *Antrocaryon* (0.88×; 38.8 × 34.46)

Astronium Jacq. Trees to 30 m tall. Leaves compound, the leaflets opposite with asymmetric bases. Infructescence terminal. Fruit, distinct in the family, a pseudodrupe surrounded by five wings, to 4 cm in length, brown when mature, typically present when the tree is defoliated. Distribution: Southern Mexico to Bolivia.

Anacardiaceae - *Astronium* (1.21×; 30.09 × 32.37)

Spondias L. Trees to 30 m tall, with very cylindrical trunks lacking buttress roots. Leaves compound, the leaflets opposite or subopposite, veins translucent, the base weakly asymmetrical. Infructescence terminal. Fruit a fleshy drupe to 5 cm long, yellow or red when mature, with rich flavor. Seeds fibrous, corky, one per fruit. Many species with edible fruits. Cultivated in tropical Africa. Distribution: Tropical America and the Old World.

Anacardiaceae - *Spondias* (1.46×; 25.51 × 16.09)

Anacardiaceae - *Spondias* (1.74×; 29.82 × 21.1)

Anacardiaceae - *Spondias* (0.86×; 45.31 × 21.48)

Anacardiaceae - *Spondias* (0.94×; 32.7 × 30.67)

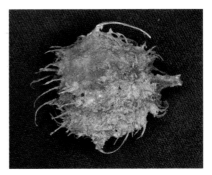

Anacardiaceae - *Spondias* (1.07×; 34.26 × 31.19)

Tapirira Aubl. Shrubs and trees to 30 m tall, typically with characteristic buttress roots at base of cylindrical trunk. Leaves compound, the leaflets opposite with asymmetric bases. Infructescence axillary. Fruit a drupe to 1.5 cm long, black when mature. Seeds one per fruit. Distribution: Tropical America.

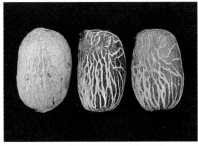

Anacardiaceae - *Tapirira* (2.33×; 12.09 × 7.95)

ANNONACEAE

Trees and shrubs, rarely woody lianas. Trunk rarely with aerial roots or resin, the inner bark aromatic with net-like venation pattern. Leaves simple, alternate, entire, pinnately veined, typically arranged distichously along the branch, aromatic. Infructescence terminal or axillary, sometimes borne from the trunk. Fruit syncarpous or apocarpous, the monocarps dehiscent or indehiscent. Seeds with ruminate endosperm. Often confused with the Lauraceae when sterile. However, the Annonaceae can be differentiated by strong bark that peels in long strips with the removal of a leaf or stem, and a monopodial branching pattern. The wood of many genera is used in rural construction and the strong bark is often used to tie and carry cargo. Many species produce edible fruit.

Anaxagorea A. St.-Hil. Shrubs and trees to 12 m tall. Leaves with characteristic tomentose pubescence, the hairs simple to stellate, and anastomosing venation most conspicuous on the abaxial laminar surface. Such characteristics can vary between individuals of the same species. Infructescence axillary

Annonaceae - Ananxagorea (2.22x; 14.81 × 7.81)

or borne from the stems. Fruit apocarpous, with up to 30 monocarps that are elongated, weakly curved, to 3 cm long, shortly stipitate and dehiscent, yellow-reddish when mature. Seeds 1–2 per monocarp, very distinct in color, brilliance, texture, and form. Distribution: Costa Rica to Peru, also in the Old World tropics.

Annona L. Primarily trees to 30 m tall, some species shrubs or lianas. Leaves with simple hairs, sometimes stellate. Infructescence axillary or terminal. Fruit syncarpous, to 25 cm diameter, generally yellow when mature, formed by numerous weakly or completely fused carpels, sometimes with small spine-like projections. Seeds one per carpel, surrounded by a fleshy, white, edible pulp. Similar to the genus Roll-

Annonaceae - Annona (1.71x; 17.04 × 10.19)

Annonaceae - Annona (4.36x; 8.03 × 4.66)

inia . Many species cultivated for their edible fruits. Distribution: Tropical America and the Old World, with many species introduced to Asia, Africa, and subtropical regions.

Cremastosperma R. E. Fries. Trees to 8 m tall. Leaves with simple hairs. Infructescence axillary or borne from the stems. Fruit apocarpous with up to 16 stipitate monocarps, to 1 cm long, black when mature. Seeds one per monocarp, with ruminate endosperm. Closely related to the genus *Guatteria*. Distribution: Peru.

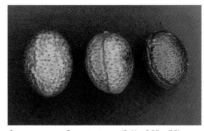

Annonaceae - Cremastosperma (2.41x; 9.25 × 7.5)

Annonaceae - Cremastosperma (3.3x; 9.1 × 6.67)

Cymbopetalum Benth. Shrubs to 4 m tall. Leaves and stems with simple hairs. Fruit apocarpous, with up to 10 weakly curved monocarps, to 5 cm long, yellow-reddish and dehiscent along one side when mature. The seeds various per monocarp, covered by an aril, with ruminate endosperm. Distribution: Northern Mexico to Bolivia and southeastern Brazil, usually in moist forest below 1000 m elevation.

Annonaceae - Cymbopetalum (2.22x; 12.62 × 8.48)

Diclinanona Diels. Trees to 20 m tall. Leaves usually large and coriaceous. Fruit apocarpous, the few monocarps larger than any other in the family, to 7 cm long, reddish to black when mature, the exocarp subwoody. Seeds usually one per monocarp. Distribution: Colombia, Venezuela, Brazil, and northeastern Peru.

Annonaceae - Diclinanona (1.21×; 32.81 × 20.35)

Duguetia A. St.-Hil. Trees to 8 m tall or shrubs. Leaves with stellate hairs or conspicuous scales. Infructescence usually borne from the stems, sometimes from the trunks. Some species produce sprawling inflorescences extending to a meter or more from the base of the trunk, with flowers and fruits that develop under the leaf litter of the forest floor. Fruit pseudosyncarpous, with numerous apiculate monocarps held closely together but not fused, to 14 cm diameter, brown and loosely separated when mature. Seeds one per monocarp, the endosperm weakly ruminate. Distribution: Brazil, Peru, and Suriname.

Annonaceae - Duguetia (1.77×; 21.2 × 8.68)

Fusaea (Baill.) Saff. Typically shrubs, but sometimes becoming trees to 20 m tall. Leaves with simple hairs and conspicuously anastomosing venation. Infructescence terminal. Fruit syncarpous, to 9 cm diameter, the carpels completely fused, green or brown when mature. Seeds numerous per fruit, the endosperm ruminate. Similar to *Annona* but differentiated by the presence of a ring at the base of the fruit formed from part of the receptacle. Distribution: Widespread in tropical America.

Annonaceae - Fusaea (2.39×; 19.12 × 9.82)

Guatteria R. & P. Trees to 30 m tall. Leaves with pubescence formed from simple hairs. Infructescence axillary or borne from the stems. Fruit apocarpous with up to 40 stipitate monocarps, to 1.5 cm long,

Annonaceae - Guatteria (3.22×; 11.25 × 6.12)

Annonaceae - Guatteria (3.89×; 13.99 × 5.41)

Annonaceae - Guatteria (2.29×; 8.94 × 5.51)

black when mature, the stipes remaining red or cherry red. Seeds one per monocarp with ruminate endosperm. Distribution: Mexico to Bolivia.

Klarobelia Chatrou. Shrubs or trees to 10 m tall. Infructescence terminal. Fruit apocarpous with up to 40 long-stipitate monocarps, to 1.5 cm long, yellow, red, or later black when mature. Seeds one per monocarp, ruminate, the surface lightly striate. Related to *Malmea*. Distribution: Lowland Amazon.

Annonaceae - *Klarobelia* (1.58x; 19.26 × 10.94)

Malmea R. E. Fries. Trees to 25 m tall. Infructescence axillary or borne from the stems. Fruit apocarpous with up to 60 stipitate monocarps to 2 cm long, red when mature; the stipes much longer than in other genera, red when mature. Seeds one per monocarp, the endosperm ruminate. Related to *Klarobelia*. Distribution: Lowland Amazon and southern Central America.

Annonaceae - *Malmea* (2.26x; 19.11 × 11.96)

Onychopetalum R.E. Fries. Trees to 28 m tall. Distinct from other genera on the basis of much thicker leaves with weaker venation. Some species exude red resin from the trunk. Fruit with one monocarp, to 6 cm long, deeply red or approaching black when mature, the mesocarp yellow, very aromatic. Seeds 1–4 per monocarp, the endosperm ruminate. Distribution: Northern Brazil to Peru.

Annonaceae - *Onychopetalum* (1.4x; 20.98 × 20.98)

Oxandra A. Rich. Trees to 28 m tall. Leaves of some species pubescent with simple hairs or with inconspicuous venation. Infructescence axillary and borne from the stems. Fruit apocarpous, with 1–6 stipitate monocarps, to 1.5 cm long, typically black when mature. Seeds one per monocarp, the endosperm ruminate. Distribution: Brazil and Peru.

Annonaceae - *Oxandra* (2.14x; 18.86 × 12.13)

Annonaceae - *Oxandra* (2.92x; 12.46 × 9.02)

Porcelia R. & P. Trees to 30 m tall. Leaves coriaceous with curved midvein. Fruit apocarpous, with 1–5 sessile monocarps, to 8 cm long, green when mature, the exocarp to 1 cm thick. Seeds flat, several per monocarp. Distribution: Brazil, Bolivia, and Peru.

Annonaceae - *Porcelia* (1.38×; 34.69 × 19)

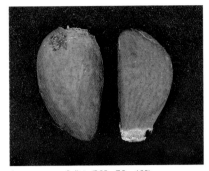

Annonaceae - *Rollinia* (5.25×; 7.3 × 4.29)

Pseudoxandra R. E. Fries. Trees to 7 m tall. Leaves with very fine anastomosing venation and pubescence of simple hairs. Infructescence axillary and borne from the stems. Fruit apocarpous, with up to 10 stipitate monocarps, to 1 cm long, black when mature. Seeds one per monocarp, the endosperm ruminate; the monocarps and seeds similar to *Guatteria* and *Oxandra*. Distribution: Venezuela and Guianas to Peru and southern Brazil.

Ruizodendron R. E. Fries. Trees to 30 m tall. Leaves glaucous abaxially. Fruit apocarpous, with up to 13 stipitate, asymmetric monocarps, to 2 cm long, green to black when mature. Seeds one per monocarp, transversely positioned, the endosperm ruminate. Monospecific genus common in flooded forests. Distribution: Bolivia, Brazil, and Peru.

Annonaceae - *Pseudoxandra* (2.29×; 10 × 10.65)

Annonaceae - *Ruizodendron* (2.03×; 16.15 × 12.45)

Rollinia A. St.-Hil. Trees to 35 m tall or shrubs. Leaves and apex of stem with dense pubescence of simple, short hairs. Fruit syncarpous, the carpels fused, sometimes apiculate, to 7 cm diameter, yellow when mature. Seeds numerous per fruit, embedded in a fleshy, edible, sometimes aromatic mesocarp. Cultivated for edible fruits. Distribution: Mexico to Bolivia.

Trigynaea Slechtend. Trees to 10 m tall. Fruit apocarpous with two short stipitate monocarps, to 3 cm long, brown and opening irregularly to expose the white, mealy mesocarp when mature, arising from somewhat woody exocarps. Seeds various per monocarp. Distribution: Venezuela and Guianas to Bolivia and Peru.

Annonaceae - *Rollinia* (2.35×; 16.08 × 8.47)

Annonaceae - *Trigynaea* (1.54×; 19.66 × 9.32)

Unonopsis R. E. Fries. Trees to 30 m tall, sometimes shrubs. Leaves with simple pubescence. Infructescence borne from the stems. Fruit apocarpous, rarely with more than 20 round, stipitate monocarps, to 2 cm long, orange to black when mature. Seeds 1–4 per monocarp, the endosperm ruminate; the size and form of the seed depends on their quantity and the size of individual monocarp. Distribution: Costa Rica to Bolivia.

Annonaceae - *Unonopsis* (2.52x; 13.07 × 13.24)

Xylopia L. Trees to 25 m tall, some species with aerial roots. Leaves usually narrow with inconspicuous venation and pubescence of simple hairs. Infructescence axillary, borne from the stems, or sometimes the trunk. Fruit apocarpous with up to 25 dehiscent, elongated monocarps to 7 cm long, red externally when mature, internally pink, weakly contracted around each seed. Seeds various per monocarp and surrounded by a thin aril. Distribution: Guatemala to Bolivia, also in the Old World tropics.

Annonaceae - *Xylopia* (3.58x; 10.38 × 5.27)

APIACEAE

Aquatic or terrestrial herbs, rarely shrubs. Plants often aromatic. Leaves simple or compound, alternate or opposite, rather variable, and often deeply lobed or dissected. Infructescence axillary or terminal. Fruit a schizocarp.

Eryngium L. Terrestrial herbs to 40 cm tall. Leaves mostly basal in the form of a rosette, the base sheathing the stem, the margin serrate to spinose. Infructescence terminal, spinose. Fruit to 0.5 cm long, brown when mature. Seeds many per fruit. Plant very aromatic. Sometimes cultivated, the leaves used as a spice like cilantro to flavor food. Distribution: Widely distributed in the tropics and southern temperate zones.

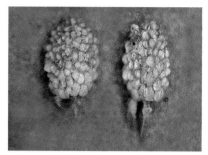

Apiaceae - *Eryngium* (16.83x; 1.6 × 1.08)

APOCYNACEAE

Trees, shrubs, and lianas. Leaves simple, entire, generally opposite, alternate, or sometimes verticillate. Some genera with foliar glands. Plants usually with white latex, sometimes transparent. Fruit a follicle, usually in pairs, with winged or arillate seeds, a few genera with berries or capsules. Many of the lianas are easily confused with the Asclepiadaceae and some authors combine the two families.

Aspidosperma Mart. et Zucc. Trees to 40 m tall. Trunks fenestrate in some species. Leaves alternate, glabrous, or pubescent. Latex absent from trunk, transparent or yellow from stems, reddish in a few species. Infructescence terminal. Fruit a follicle, solitary or in pairs, woody, smooth or rugose, glabrous, or pubescent, to 15 cm long, brown or black when mature. Seeds numerous per fruit, completely surrounded by a membranous, papyraceous wing; funicle persistent. Many species used for construction, posts, and carpentry. Bark medicinal. Distribution: Mexico, Guianas, Venezuela to Bolivia.

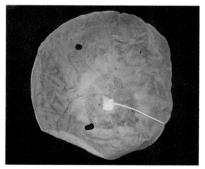

Apocynaceae - *Aspidosperma* (0.79x; 62.68 × 61.07)

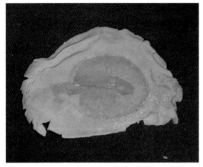

Apocynaceae - *Aspidosperma* (0.95×; 58.55 × 37.99)

Apocynaceae - *Forsteronia* (0.77×; 19.04 × 1.92)

Geissospermum F. Allen. Trees to 25 m tall, with fenestrate trunk. Leaves alternate. Infructescence terminal or axillary and seeming to be borne from the stems. Fruit a berry to 3.5 cm long, yellow when mature. Seeds 6–10 per fruit, surrounded by a milky, fleshy mesocarp. Distribution: Guianas, Brazil, Bolivia to Peru.

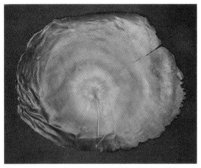

Apocynaceae - *Aspidosperma* (0.57×; 101.67 × 80)

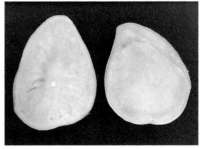

Apocynaceae - *Geissospermum* (2.86×; 13.41 × 10.37)

Himatanthus Willd. Trees to 20 m tall, often growing in disturbed areas. Leaves alternate, spiral, clustered at branch apex. Latex white, abundant. Fruit a follicle, in pairs, to 35 cm long, dark brown when mature. Seeds numerous per fruit, completely surrounded by a wing of the same color. Wood is harvested for lumber. Latex, roots and fruit reported to have medicinal properties. Distribution: Guianas, Brazil, Bolivia to Peru.

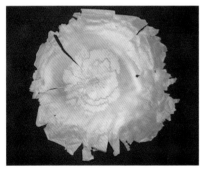

Apocynaceae - *Aspidosperma* (0.47×; 106.44 × 103.95)

Forsteronia G.F.W. Mey. Lianas. Leaves opposite; some species with foliar glands. Latex white. Infructescence axillary or terminal. Fruit a pair of elongate, linear, cylindrical follicles to 20 cm long, brown when mature. Seeds numerous per fruit, plumose, the hairs golden-yellow. Distribution: Guianas, Brazil, Bolivia to Peru.

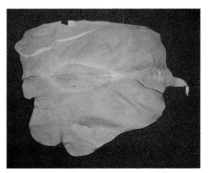

Apocynaceae - *Himatanthus* (1×; 52.7 × 44.87)

Lacmellea Karst. Shrubs and trees to 15 m tall. Leaves opposite. Latex white, abundant. Infructescence axillary or terminal. Calyx persistent. Fruit a round berry, to 2 cm diameter, yellow when mature. Seeds one or rarely up to three per fruit surrounded by a papryceous covering that is very fragile when dry. Distribution: Guianas, Brazil to Peru.

Apocynaceae - *Lacmellea* (3.45×; 7.76 × 5.43)

Macoubea Aubl. Trees to 30 m tall. Leaves opposite. Latex white, abundant. Infructescence terminal. Fruit a dry globose berry, to 12 cm diameter, brown when mature, the exocarp woody, mesocarp yellow with white latex. Seeds 35–45 per fruit, loose in the locule at full maturity. Distribution: Guianas to Brazil and Peru.

Apocynaceae - *Macoubea* (2.22×; 18.38 × 6.95)

Mucoa Zarucchi. Trees to 30 m tall. Leaves opposite with fine tertiary venation almost perpendicular to the secondary veins. Latex white, abundant. Infructescence axillary. Fruit a berry to 9 cm diameter, brown when mature, the exocarp woody. Seeds many per fruit. Distribution: Colombia, Brazil, and northeastern Peru.

Apocynaceae - *Mucoa* (4.28×; 10.17 × 4.95)

Odontadenia Benth. Lianas. Leaves opposite. Latex white, abundant. Infructescence axillary. Fruit a woody follicle, solitary or in pairs, to 20 cm long, brown when mature. Seeds numerous per fruit, with plumose, golden-brown hairs. Distribution: Costa Rica and Trinidad to Peru.

Apocynaceae - *Odontadenia* (0.67×; 20.48 × 3.02)

Pacouria Aubl. Lianas. Leaves opposite. Latex white. Infructescence axillary or terminal. Fruit a globose berry, to 15 cm diameter, yellow or orange when mature, exocarp thick with abundant white latex, mesocarp yellow, fleshy, fibrous, aromatic, sweet-sour taste. Seeds numerous per fruit embedded in the mesocarp, with a papyraceous covering derived from the cuticle that defoliates when dry. Distribution: Surinam and Guianas to Peru.

Apocynaceae - *Pacouria* (1.5×; 20.49 × 20)

Parahancornia Ducke. Trees to 30 m tall. Leaves opposite. Latex white, abundant. Infructescence axillary or terminal. Fruit a berry, globose, to 5 cm diameter, brown when mature, the exocarp thick, subwoody. Seeds various per fruit. Distribution: Colombia, Brazil, and northeastern Peru.

Apocynaceae - *Parahancornia* (2.31x; 13.26 × 6.7)

Prestonia R. Br. Lianas. Leaves opposite. Fruit a follicle, in pairs, to 20 cm long, light brown when mature. Seeds numerous per fruit with plumose, golden-brown hairs. Plants generally pubescent. Distribution: Southern Mexico and Guianas to Bolivia and Peru.

Apocynaceae - *Prestonia* (2.07x; 10.46 × 2.14)

Rauvolfia L. Trees 30 m tall or treelets. Leaves 3- to 5-verticillate. Fruit a berry to 2 cm diameter, green-yellow to purple when mature, mesocarp white, with

Apocynaceae - *Rauvolfia* (2.14x; 20.25 × 10.79)

sticky, white latex and strong odor. Seeds one per fruit. Wood used for lumber. Leaves and roots used for medicinal properties. Distribution: Brazil, Peru, Bolivia, and the Old World tropics.

Apocynaceae - *Rauvolfia* (1.93x; 17.8 × 19.29)

Tabernaemontana L. Trees to 30 m tall or shrubs. Leaves opposite, variable in size. Latex white, abundant. Fruit a follicle, in pairs, to 4 cm long, green, brown, or red externally when mature, the inner walls pinkish-red, and in some species the exocarp lightly spinose. Seeds various per fruit, covered almost completely by an orange aril. Distribution: Costa Rica and Trinidad to Peru.

Apocynaceae - *Tabernaemontana* (2.05x; 13.87 × 6.8)

AQUIFOLIACEAE

Trees or shrubs. The leaves black when dry. Cosmopolitan distribution in temperate and tropical regions with only one genus and few species in the Amazonian region.

Ilex L. Shrubs or trees to 25 m tall. Leaves simple, alternate, weakly coriaceous. Infructescence axillary. Fruit a berry to 0.5 cm diameter, yellow or red when mature. Seeds 3–5, rarely more than eight per fruit. Various species are used for their wood in construction and carpentry, others are used as ornamentals. Distribution: Americas and Old World, temperate and tropical zones.

Aquifoliaceae - *Ilex* (4.76×; 10.27 × 3.4)

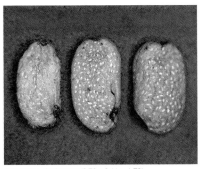

Araceae - *Anthurium* (9.79×; 3.11 × 1.73)

ARACEAE

Herbs, primarily growing as epiphytes and hemi-epiphytes, some terrestrial and a few aquatic. Leaves simple, rarely compound, alternate, coriaceous, succulent, sometimes with perforations and lobes, sometimes with strong morphological dimorphism when juveniles, the petioles succulent and often elongated, their bases sheathing the stem. Some genera with transparent, mucose sap that causes skin irritation. Infructescence terminal, consisting of a spadix of flowers that is often covered by the hood-like structure known as the spathe. Fruit a berry with numerous seeds, maturing on the spadix to produce a compound, multiple fruiting structure. The majority of genera prefer humid habitats.

Araceae - *Anthurium* (6.19×; 5.5 × 2.46)

Anthurium Schott. Epiphytic and terrestrial herbs, sometimes lianas. Leaves simple, alternate, lanceolate, and entire, but sometimes cordate, lobed, or divided into multiple segments. The genus can often be recognized by the presence of anastomosing venation. Spathe often reduced, not covering the spadix. Infructescence erect or pendent, sometimes very large up to 90 cm. Fruits arranged in groups along the spadix, of various colors when mature, often red. Seeds 1–3 per fruit. Many species with a mass of roots hanging below the plant, and often associated with ant gardens. Numerous species cultivated as ornamentals. Distribution: Southern Mexico and Central America, Trinidad and Tobago, to Bolivia and Peru.

Dieffenbachia Schott. Succulent terrestrial herbs sometimes to 2 m tall. Leaves entire, many times with white spots on the adaxial side of the leaf. Sap caustic, malodorous. Spathe persistent, completely covering and enclosing the spadix. Fruits clustered lightly around the spadix, green-yellow, opening irregularly when mature, exposing a white, mealy mesocarp. Seeds two per fruit, with orange aril. Many species cultivated as ornamentals. Distribution: Ecuador, Brazil, and Peru.

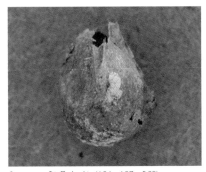

Araceae - *Dieffenbachia* (6.24×; 6.27 × 5.39)

Dracontium L. Terrestrial herbs to 2 m tall. Leaf solitary, deeply lobed, annual, disappearing after fruiting; the veins anastomosing; the petiole very large, emerging from the soil, brown with variegation, often reminding one of snake skin. Many spe-

Araceae - *Anthurium* (8.31×; 4.11 × 3.36)

cies have inflorescences with fetid odor during anthesis. Spathe brown, covering the spadix. Infructescence green, emerging from the soil, consisting of numerous fruits tightly clustered around the spadix to form a compact structure that appears almost syncarpous. Fruits conical, green when mature, the mesocarp white to transparent, sticky. Seeds 1–2 per fruit. Distribution: Brazil and Peru.

Araceae - *Dracontium* (2.05×; 8.92 × 7.42)

Heteropsis Kunth. Hemiepiphyte climbers. Roots to 25 m long, hanging from the branches of trees, very strong, resistant. Leaves coriaceous, entire, veins anastomosing; the petiole very short, not succulent, distinct within the family. Spathe reduced, caducous. Infructescence consisting of numerous fruits clustered loosely or tightly around the spadix to form a compact structure that appears almost syncarpous. Fruits variable in size and shape, yellow or orange when mature, the mesocarp transparent to white. Seeds one per fruit. Roots often used as rope and in handicrafts. Distribution: Guianas, Brazil to Peru.

Araceae - *Heteropsis* (2.46×; 18.32 × 8.06)

Araceae - *Heteropsis* (3.37×; 10.66 × 7.83)

Monstera Adans. Hemiepiphitic climbers. Leaves entire in juvenile state, growing very tightly on tree trunks, generally with orifices and deep lobes or sometimes entire in adult state. Spathe larger than the spadix, caducous. Infructescence consisting of numerous fruits clustered loosely around the spadix. Fruits conical, yellow or orange when mature, the mesocarp white. Seeds one per fruit. Distribution: Mexico to Peru.

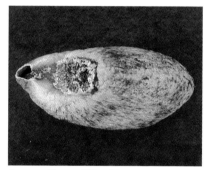

Araceae - *Monstera* (5.95×; 10.34 × 4.64)

Araceae - *Monstera* (5.67×; 9.93 × 4.53)

Philodendron Schott. Hemiepiphytic or epiphytic climbers, rarely terrestrial. Leaves cordate, entire, or rarely trilobed to multisegmented, with strong heterophylly. Some species with lanceolate leaves and with very short or winged petioles. Spathe persistent, generally reddish, completely surrounding the spadix. In-

Araceae - *Philodendron* (12.91×; 1.68 × 1.13)

fructescence consisting of numerous fruits clustered around the spadix to form a compact structure. Fruit cream-yellow, orange, or red when mature. Seeds very small, usually many per fruit; a few species with one seed per fruit. Distribution: Mexico to Peru.

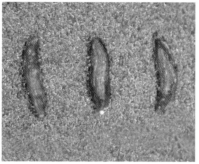

Araceae - *Philodendron* (26.03x; 1.04 × 0.27)

Araceae - *Philodendron* (6.56x; 9.06 × 4.35)

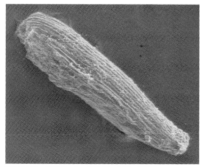

Araceae - *Philodendron* (63.49x; 1.11 × 0.29)

Syngonium Schott. Hemiepiphytic climbers. Leaves 3- to 5-lobed. Veins anastomosing. Spathe completely enveloping spadix, caducous, yellow when mature. Infructescence consisting of numerous fruits clustered loosely around the spadix. Fruit with thin, brownish exocarp when mature, the mesocarp white, sweet. Seeds many per fruit. Distribution: Brazil, Ecuador, Peru, and Bolivia.

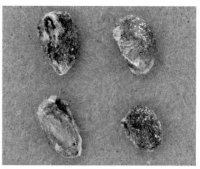

Araceae - *Syngonium* (2.96x; 7.11 × 4.64)

Xanthosoma Schott. Terrestrial herbs to 3 m tall. Leaves sagittate with large basal lobes, the veins conspicuously anastomosing, irregular. Spathe caducous. Infructescence consisting of fruits clustered loosely around the spadix. Fruit yellow when mature. Distribution: Colombia, Ecuador, Peru, and Bolivia.

Araceae - *Xanthosoma* (13.97x; 1.63 × 1.02)

ARALIACEAE

Shrubs or trees, rarely hemiepiphytes. Leaves alternate, simple, bipinnate, or palmately compound. Stipules large and persistent. Infructescence terminal. Fruit a berry. Cosmopolitan family of temperate and tropical regions.

Dendropanax Decne. et Planchón. Trees to 6 m tall. Leaves simple, entire when adults, or sometimes trilobed in juvenile form. Leaves and petioles variable in size, somewhat trinerved from the base. Styles persistent. Fruit a berry, to 0.7 cm diameter, black when mature. Seeds 5–7 per fruit. Plants used for honey production, the wood used in carpentry, plywood, and

Araliaceae - *Dendropanax* (6.83x; 3.88 × 2.29)

matchsticks. Leaves and roots are medicinal. Distribution: New World and the Old World tropics.

Schefflera J.R. Forst. & G. Forst. Trees to 20 m tall. Trunk with a strongly aromatic sap. Leaves palmately compound, leaflets up to 12, clustered at the apices of the branches, sometimes with lobes, the petioles elongated and variable in size. Stigma persistent on fruit. Fruit a berry, to 2 cm diameter, purple or black when mature, longitudinally flattened. Some species with dry fruit composed of two parts, but the majority of species with a fleshy berry of 3–9 parts. Seeds 2–9 per fruit. Wood used in carpentry, to make matchsticks, and sometimes as a substitute for balsa wood. The petioles are sometimes used to make toys, as well as cages for birds. Leaves, resin, and bark medicinal. Distribution: New World and the Old World tropics.

Araliaceae - *Schefflera* (3.41×; 5.68 × 8.48)

ARECACEAE

Family diverse and variable. Appearing like trees or shrubs, herbaceous to woody, without branches, rarely lianas. Stems solitary or clustered or rarely subterranean, often with spines. Leaves usually clustered near stem apex, simple or compound, pinnate or palmate, the petiole base sheathing the stem, often leaving a circular scar after dehiscing. Infructescence in racemes or pseudospikes, covered initially by a bract. Fruit a drupe with one or rarely two seeds, some genera with more than two embryos per seed. Family typical of tropical and subtropical regions, and one of the most important economically.

Astrocaryum G.F.W. Mey. Palms to 10–20 m tall. Stems solitary, rarely clustered or subterranean, with conspicuous leaf scars and large, flattened, black, caducous spines. Leaf rachis to 6 m long with black flat spines. Leaves with many pinnae, abaxially glaucous, the margins with short spines. Bract to 1 m long. Infructescence to 80 cm. Fruit sessile, orange or brownish, with very small hair-like spines on the exocarp, the mesocarp fleshy, abundant, somewhat fibrous and sweet. Calyx persistent. Seeds one per fruit with three pores on the apex. Young leaves of some species are used to make handicrafts. Fruits edible. Distribution: Mexico and Guianas to Peru and Bolivia, to 650 m elevation.

Arecaceae - *Astrocaryum* (1.62×; 34.71 × 23.4)

Attalea H. B. K. Palms to 25 m tall and stem to 40 cm diameter. Stems solitary, sometimes subterranean. Leaves with numerous pinnae. Rachis to 12 m long, arched and curved in such a manner that the pinnae appear vertical in the apex. Normally, the pinnae arranged in one plane, but in some species the pinnae are disorganized. Bract to 1 m long, indurate to woody. Infructescence to 60 cm. Fruits tightly clustered, sessile, brown when mature, the exocarp woody, the mesocarp white or orange and fibrous. Seeds one per fruit with 3–4 embryos per seed. Leaves used as roofing material for houses. Fruits edible. Seeds often used in making handicrafts. Distribution: Mexico to Bolivia.

Arecaceae - *Attalea* (0.76×; 71.67 × 31.67)

Arecaceae - *Attalea* (1.09×; 45.83 × 21.07)

Bactris Jacq. ex Scop. Palms 1–20 m tall. Stems solitary or clustered forming small or large colonies, unarmed or with spines that form black or yellow rings around the stem. Leaves highly variable, pinnate, generally only entire in the apex or sometimes entire and bifid in the apex. Pinnae sometimes arranged in groups of 2–4, disorganized, with or without small spines in the margin. Some species with very acuminate pinnae. Rachis to 4 m, with or without spines. Bract generally covered with small black spines. Infructescence to 40 cm, sometimes emerging from the soil at the stem base, usually erect. Peduncle and bract armed with spines or unarmed. Fruit sessile, elongate when clustered, or round when in racemes, black, orange, yellow, or red, armed with short spines or unarmed, the mesocarp fleshy, juicy, or dry. Calyx

persistent. Seeds one per fruit, with tree pores close to the middle. Some species have edible fruits, and the wood is sometimes used to make bows. Distribution: Mexico and Guianas to Bolivia and Paraguay, to 1800 m elevation.

Chamaedorea Willd. Palms to 5 m tall. Stems solitary, rarely clustered, rarely subterranean. Leaves entire, bifid at the apex, or pinnately lobed. Rachis to 2 m long. Infructescence 26–40 cm, usually arising directly from the stem. Peduncle orange or red. Calyx persistent. Fruit red to orange or black when mature, sessile and shiny, the mesocarp green-yellow or orange. Seeds one per fruit. The flowers of some species used as a fragrance and for extraction of perfume. Distribution: Mexico to Bolivia, to 2700 m elevation.

Arecaceae - *Bactris* (1.85x; 30 × 12)

Arecaceae - *Chamaedorea* (3.75x; 10.71 × 6.89)

Arecaceae - *Bactris* (3.07x; 12.97 × 7.97)

Arecaceae - *Chamaedorea* (3.34x; 10.41 × 5.73)

Chelyocarpus Dammer. Palms to 8 m tall. Stems usually solitary, rarely growing in clusters. Leaves palmate to 1 m diameter, with many segments, each segment with two lobes, lightly glaucous below. The bract covered with a light-colored woolly pubescence on the exterior. Infructescence to 24 cm long. Fruit brown when mature. Seeds one per fruit. Natives in Colombia burn, cook, and filter the stems to extract vegetable salt that is mixed with " *ambil de tobacco*." Distribution: Colombia, Ecuador, Peru, and Brazil.

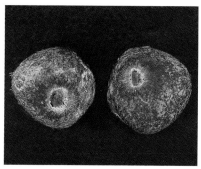

Arecaceae - *Bactris* (4.44x; 6.57 × 6.52)

Arecaceae - *Chelyocarpus* (1.61x; 18.16 × 17.76)

Desmoncus Mart. Climbing palms. Stems with spines formed from modified leaves. Leaves pinnate, the pinnae generally opposite, the terminal pinnae modified into spines. Rachis to 100 cm. Spines short, curved toward the base and angled, on the stem, leaf rachis, and on veins of the basal surface of the leaf, scattered and smaller on the bract. Infructescence 11 cm, axillary, with two bracts. Calyx persistent. Fruit sessile, yellow, orange or red, the mesocarp orange. Seeds one per fruit. Distribution: Mexico, Guianas, and Trinidad to Peru and Bolivia.

Arecaceae - *Desmoncus* (2.05x; 17.22 × 10.31)

Euterpe Mart. Palms to 25 m tall. Stems solitary or clustered, more or less smooth. Leaves to 3 m with many pairs of very slender pinnae. Infructescence to 50 cm. Calyx persistent. Fruit sessile, black, very

Arecaceae - *Euterpe* (1.57x; 13.31 × 12.57)

round, the mesocarp thin, edible. Seeds one per fruit. Characterized by a group of slender red aerial roots at the base of the trunk. Some species weakly cultivated for use of young leaves in salads and for the fruits, which are used in beverages. Trunks are used in rural construction. Leaves and roots medicinal. Distribution: Central America and Trinidad to Bolivia, to 3000 m elevation.

Geonoma Willd. Palms to 8 m tall. The stems solitary, clustered, or sometimes subterranean, with characteristic rings originating from points of leaf attachment. Leaves entire or pinnate, the apex entire or bifid. Rachis to 2 m long. Infructescence 70 cm, spicate or paniculate/branched. The peduncle and the infructescence branches green, orange, or red. Calyx persistent. Fruit sessile, black, and round or elongate. Seeds one per fruit. The leaves of some species are used as thatch. Distribution: Mexico and Guianas to Bolivia, to 1200 m elevation.

Arecaceae - *Geonoma* (4.97x; 5.79 × 5.74)

Arecaceae - *Geonoma* (3.28x; 11.1 × 9.03)

Hyospathe Mart. Palms to 1.5 m tall. Stems clustered, rarely solitary, with characteristic concentric rings. Leaves rarely entire, usually pinnate to 100 cm, the apex entire. Infructescence to 20 cm. Peduncle red, well-developed with a flattened base, the infructescence branches reddish. Calyx persistent. Fruit sessile, elongate, black. Seeds one per fruit. Distribution: Central America to Bolivia, to 2000 m elevation.

Iriartea Ruiz & Pavón. Palms up to 25 m tall. Stems solitary, the trunk with a very pronounced conspicuous swelling in the upper half, below the leaves. The aerial roots to 2 m long, dense, lenticellate, black, the young ones appearing phallic. Leaves pinnate, the pinnae triangular and disorganized, the apex entire. Rachis to 3 m. Bract to 1 m long in the form of a horn. Infructescence to 1.5 m with abundant sessile fruit, black when mature, the exocarp smooth and veined or with small canals when mature, the mesocarp white, mealy. Seeds one per fruit. The stem used in rural construction. Distribution: Nicaragua to Bolivia, to 1500 m elevation.

Arecaceae - *Lepidocaryum* (2.86x; 8.93 × 8.33)

Mauritia L. f. Palms to 25 m tall, dioecious. Stems solitary. Leaves palmate with many segments, large. The bract to 3 m long. Infructescence to 2 m long. Calyx persistent. Fruit subglobose, sessile, dark red to purple when mature, the exocarp with shiny scales, the mesocarp yellow to orange-yellow, thin, edible. Seeds usually one per fruit, sometimes two. Some species widely commercialized for their fruits. Locals eat a large weevil larvae (Curculionidae) extracted from the decomposing trunk—known as 'suri' in the Peruvian Amazon. Distribution: Trinidad, Guianas, Amazonian basin and the Orinoco basin, growing primarily in swamps, at the borders of lakes, forming dominant stands that can cover vast areas.

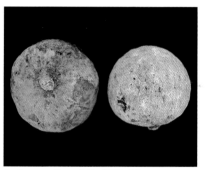

Arecaceae - *Iriartea* (1.53x; 20.28 × 20)

Iriartella H. A. Wendl. Palms to 10 m tall. Stems solitary. Aerial roots small, short. Leaves pinnate, the pinnae divided into many segments and organized in various planes or angles. Rachis to 2 m long. Calyx persistent. Fruit subglobose, sessile, yellow, irregularly splitting when mature to expose the white, mealy mesocarp. Seeds one per fruit. Distribution: Northeastern Peru, Colombia, and Brazil.

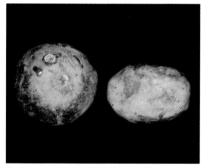

Arecaceae - *Mauritia* (0.98x; 28.6 × 29.46)

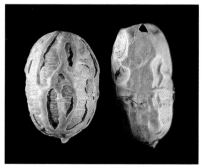

Arecaceae - *Iriartella* (3.87x; 10.63 × 7.46)

Lepidocaryum Mart. Palms to 3 m tall. Stems solitary. Leaves palmate with four segments. Fruit subglobose, sessile, dark red to dark purple to purple-black when mature, the exocarp with shiny scales, the mesocarp yellow, very thin, edible. Seeds one per fruit. Distribution: Northeastern Peru, Colombia, and Brazil.

Mauritiella Burret. Palms to 15 m tall, dioecious. Stems clustered, lower trunk and roots with conical spines. Leaves palmate, circular, deeply divided into numerous segments, the margins spinose, the abaxial surface glaucous, and the petiole extended into the lamina. Infructescence axillary. Fruit to 3.5 cm long, smaller than fruits of *Mauritia*, globose, covered with scales organized in vertical spirals, reddish when mature, the mesocarp thin and mealy. Seeds one per fruit. Distribution: South America to 1500 m elevation, growing in poorly drained to permanently inundated habitats or along river margins.

Arecaceae - *Mauritiella* (2.5x; 19.11 × 17.29)

Oenocarpus Mart. Palms to 20 m tall. Stems solitary or clustered with numerous small, slender aerial roots at the base. Leaves pinnate with many pairs of pinnae, shorter at the apex, in one plane, erect, often glaucous below. The rachis to 7 m long. Petiole with black fibers that are abundant along the margin. Bract to 1.6 m. Infructescence to 100 cm. Calyx persistent. Fruit round, sessile, black, the mesocarp thin. Seeds one per fruit. The fruits are used to prepare beverages and in some species they produce one of the finest vegetable oils in the world. The leaves are used as thatch in rural construction. The roots are reportedly medicinal. Distribution: Costa Rica to Peru and Bolivia, to 1400 m elevation.

Arecaceae - *Oenocarpus* (1.46x; 36.96 × 21.74)

Arecaceae - *Oenocarpus* (2.58x; 19.18 × 15.82)

Pholidostachys Blume. Palm to 2 m tall. Leaves pinnate, irregularly divided. Infructescence with very short peduncle. Fruit to 2 cm long, black when mature, the mesocarp strongly fibrous. Seeds one per fruit. Distribution: Northeastern Peru, Colombia, and Brazil.

Arecaceae - *Pholidostachys* (2.65x; 18.12 × 10.4)

Phytelephas Ruiz & Pavon. Palms to 15 m, dioecious. Stems solitary or clustered, short to subterranean. Leaves pinnately compound, glabrous, the raquis to 3 m long, the pinnae evenly distributed in one plane. Infructescence axillary. Peduncle large. Fruits large, to 20 cm diameter, tightly grouped in a dense cluster, tuberculate, with subconical protuberances, woody, brown when mature, the mesocarp mealy, yellow, or orange. Seeds one per fruit, extremely hard; known as vegetable ivory and used in arts and crafts and to manufacture buttons. Distribution: Panama to Bolivia.

Arecaceae - *Phytelephas* (1.17x; 47.75 × 29.82)

Socratea G. Karsten. Palms to 25 m tall. Stems solitary with long, spinose aerial roots. Leaves pinnate, the pinnae divided into many segments and organized in various planes. Rachis to 3 m long. Infructescence to 1 m long. Calyx persistent. Fruit subglobose, sessile, to 2.5 cm diameter, yellow, splitting irregularly when mature to expose the white, mealy mesocarp. Seeds one or rarely two per fruit. Distribution: Nicaragua to Bolivia, to 1000 m elevation.

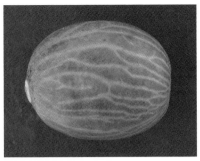

Arecaceae - *Socratea* (2.26x; 22.38 × 17.57)

Wettenia Poeppig. Palms to 15 m tall. Stems solitary or clustered with short, spinose or lenticellate roots at the base. Leaves pinnate, the pinnae asymmetrically divided into various segments. The rachis to 3 m long. Infructescence to 30 cm long, brown, the fruits tightly aggregated forming a very compact structure. Fruit conical, of different shapes and sizes, with tan to brown pubescence. Seeds one per fruit. The trunks are parted in thin boards and used in construction. The leaves are used as thatch. Distribution: Panama to Bolivia, to 800 m elevation.

Arecaceae - *Wettenia* (1.81x; 20.82 × 11.4)

ARISTOLOCHIACEAE

Lianas. Leaves alternate, simple, entire, often with a cordate base and a peculiar odor. Fruit a septicidal capsule opening in various segments. Due to similar growth habit and leaves, sometimes confused with the Passifloraceae, but easily distinguished by lack of tendrils.

Aristolochia L. Lianas. Bark often soft, corky. Leaves often cordate, the secondary venation conspicuous and somewhat parallel. Infructescence axillary. Pedicel splitting as fruit opens. Fruit a septicidal capsule with six valves opening longitudinally, to 15 cm long, brown when mature, sometimes remaining united to form a basket-like structure. Seeds winged, numerous per fruit. Distribution: Southern United States through Mexico and Central America to Peru and Bolivia, also in Asia and Africa.

Aristolochiaceae - *Aristolochia* (3.6x; 11.29 × 9.03)

ASCLEPIADACEAE

Lianas. Leaves simple, opposite, entire, generally with glands at the base. Latex white, abundant. Fruit a follicle with numerous seeds, each with plumose hairs. Easily confused with lianas of the Apocynaceae and treated by some authors as part of that family.

Marsdenia R. Br. Lianas. Leaves often deciduous when fruit is present. Fruit a follicle, the pericarp smooth, indurate, somewhat woody, to 25 cm long, yellow-green when immature, brown when mature. Seeds numerous per fruit, plumose, the hairs white. Distribution: Mexico to Argentina.

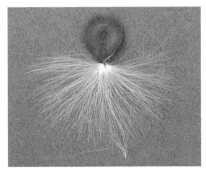

Asclepidaceae - *Marsdenia* (1.35x; 12.11 × 10)

ASTERACEAE

One of the largest and most variable families of flowering plants. Herbs, vines, shrubs, or trees to 30 m, rarely epiphytes. The stems often hollow with a spongy pith. Leaves simple, alternate, opposite, or verticillate, rarely compound. Infructescence a compound head (capitulas) arising from the peduncle, with flowers positioned on a receptacle surrounded by involucral bracts. Fruit an achene, often with persistent pappus. Family divided into 13 tribes. Plants sometimes with latex, spines, or a strong odor or fragrance.

Clibadium F. Allam. ex L. Herbs or shrubs to 2 m tall. Leaves simple, opposite, sometimes weakly trinerved, and with the margin smoothly serrate. Infructescence terminal. Fruit an achene with persistent pappus. If lacking pappus, then giving the aspect of a small, fleshy drupe. Plants sometimes aromatic. Distribution: Costa Rica to Paraguay.

Asteraceae - *Clibadium* (15.87×; 1.39 × 1.13)

Conyza Less. Herbs to 1.5 m tall. Leaves simple, alternate, lanceolate, margin weakly serrate. Infructescence axillary. Fruit an achene. Pappus of white bristleform hairs. Distribution: Cosmopolitan, better represented in the temperate regions.

Asteraceae - *Conyza* (7.51×; 1.28 × 0.37)

Asteraceae - *Conyza* (9.63×; 1.23 × 0.34)

Erechtites Raf. Herbs to 2 m tall. Leaves simple, alternate, sessile or subsessile, generally lanceolate, margin dentate or weakly lobed. Infructescence axillary and terminal. Involucral bracts 1.2 cm. Fruit an achene. Pappus of soft bristles, white to pink falling easily at maturity (and thus lacking in the following seed image). Distribution: Canada to Argentina, also in central Europe and Hawaii.

Asteraceae - *Erechtites* (15.24×; 2.47 × 0.59)

Mikania Willd. Herbaceous vines. Leaves simple, opposite, ovate to cordate, usually 3- to 5-nerved from base of lamina. Infructescence axillary or terminal. Fruit an achene, dark brown when mature. Plants sometimes aromatic. Some species with white latex in the stems. Distribution: Mexico to Bolivia and Peru.

Asteraceae - *Mikania* (4.44×; 3.76 × 0.52)

Asteraceae - *Mikania* (5.45×; 4.08 × 0.45)

Piptocarpha R. Br. Vines or shrubs, rarely trees. Leaves simple, alternate, entire, abaxially glaucous or with dense pubescence of stellate hairs. Infructescence axillary. Fruit white when mature, 10-sided. Pappus white, rarely absent. Vegetatively sometimes confused with the genus *Solanum* (Solanaceae). Distribution: Mexico to Bolivia and Peru.

Asteraceae - *Piptocarpha* (5.66x; 2.97 × 0.95)

Porophyllum Guett. Herbs to 1.5 m tall. Leaves alternate, rarely opposite, very thin, with undulating margin and inconspicuous pinnate venation, sometimes with glands on the lamina. Infructescence axillary. Bracts covering the mature seed but not the persistent stigma. Fruit an achene, brown when mature. Plants strongly aromatic and frequently growing in high rocky habitats. Distribution: United States, Mexico, Central America, and the Antilles, to Argentina.

Asteraceae - *Porophyllum* (3.09x; 7.77 × 0.55)

Pseudoelephantopus Rohr. Herbs to 0.4 m tall. Leaves alternate, sessile, sheathing the stem, the margin weakly undulate to serrate, some species glaucous abaxially. Infructescence terminal. Fruit an achene, brown when mature. Plants pubescent throughout and common in disturbed habitats. Distribution: Costa Rica and the Antilles to Argentina.

Asteraceae - *Pseudoelephantopus* (7.72x; 3.19 × 0.64)

Vernonia Schreber. Herbs and shrubs to 2 m tall, rarely lianas or trees to 20 m tall. Leaves alternate, entire or weakly dentate, sessile or short-petiolate, with pinnate venation. Infructescence axillary or terminal. Bracts to 0.6 cm. Fruit an achene, brown when mature. Pappus bristles white. Some species pubescent throughout. Frequently in areas of full sun, such as at the margin of rivers and forests, and in forest gaps. Distribution: Temperate and tropical regions of the Americas, Africa, and Asia.

Asteraceae - *Vernonia* (5.4x; 2.24 × 0.78)

Wulffia Necker ex Cass. Lianas or climbing herbs to 2 m tall. Stems square. Leaves opposite, short pubescent, sometimes asperous, margin entire or serrate, the venation trinerved or pinnately nerved. Infructescence terminal or axillary with orange bracts.

Asteraceae - *Wulffia* (7.25x; 4.32 × 1.94)

Fruit an achene, black when mature. Plants very aromatic, common in secondary forest. Distribution: Panama to Paraguay.

cent. Seeds small, many per fruit. Distribution: Mexico to Peru and Bolivia, but more speciose in the Old World tropics.

BALANAPHORACEAE

Herbs, parasitic on roots of other plants. Leaves absent. Infructescence and peduncle visible. Often easily confused with fungus.

Ombrophytum Poepp. ex Endl. Herbs, parasitic. Fruiting body to 10 cm, red when mature, the fertile terminus to about 6 cm long. Seeds many per fruit. Distribution: Peru, Bolivia, and Argentina, to 3500 m elevation.

Balanaphoraceae - *Ombrophytum* (13.86×; 1.57 × 1.21)

BEGONIACEAE

Small family of herbs and vines, rarely shrubs. Leaves alternate, simple, asymmetric, the margin serrate. Stipules present. Stems succulent with many nodes giving rise to adventitious roots in climbing species.

Begonia L. Herbs and vines, rarely shrubs. Leaves succulent, weakly serrate, thin when dry. Infructescence terminal. Fruit a capsule, to 3 cm, brown when mature, with three wings, one elongated and two reduced, reticulately nerved and somewhat translu-

Begoniaceae - *Begonia* (88.89×; 0.53 × 0.22)

BIGNONIACEAE

Family composed primarily of lianas, with some trees and shrubs. Tendrils simple to terminally 3-branched. Some genera with glands. Leaves compound, rarely simple, opposite; the majority of lianas trifoliate, often with the third leaflet modified into a tendril, and the trees with palmately or bipinnately compound leaves; few genera with simple leaves, rarely alternate. Infructescence axillary or terminal, sometimes borne from the trunk. Fruit a bivalved capsule, septicidal or loculicidal leaving a septum, rarely a berry. Seeds winged, numerous per fruit. Some species used for ornamental purposes and others for their high quality, dense wood.

Anemopaegma Mart. ex. Meisn. Lianas. Tendrils simple or 3-branched. Leaves compound, opposite, bifoliate or trifoliate. Infructescence axillary. Fruit a capsule, to 15 cm long, brown when mature. Seeds winged, many per fruit. Distribution: Belize to Peru and Bolivia.

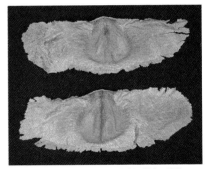

Bignoniaceae - *Anemopaegma* (1.06×; 56.3 × 19.9)

Arrabidaea DC. Lianas. Some species with square stems. Leaves compound, opposite, trifoliate, or bifoliate, with the third leaflet modified into a simple tendril. One species with simple leaves, and another with biternately compound leaves. Glands often pres-

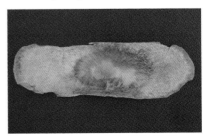

Bignoniaceae - *Arrabidaea* (0.86×; 72.22 × 20.62)

ent between the petioles. Infructescence terminal or axillary. Fruit a bivalved capsule, to 30 cm long, smooth or verrucose, with a prominent midvein, brown or black when mature. Seeds winged, numerous per fruit. Distribution: Mexico to Argentina and Paraguay.

Clytostoma Miers ex Bur. Lianas. Stems often square. Leaves compound, opposite, trifoliate or bifoliate with third leaflet modified into a simple tendril. Infructescence terminal or axillary. Fruit a bivalved capsule, to 10 cm long, rarely to 20 cm, often warty, the protuberances to 1 cm long, the valves dehiscing in two parts when dry, brown when mature. Seeds winged, numerous per fruit, in two rows. Axillary pseudostipules present. Similar to *Cydista*, differs in having warty fruit. Distribution: Mexico to Argentina and Paraguay.

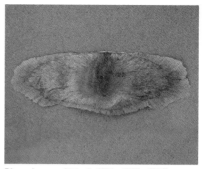

Bignoniaceae - *Distictella* (0.91x; 67.09 × 22.56)

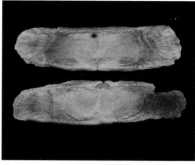

Bignoniaceae - *Distictella* (1.36x; 44.05 × 11.31)

Jacaranda Juss. Trees to 35 m, common in secondary forest habitats. Leaves bipinnately compound, opposite, to 170 cm long. In Brazil some species with leaves pinnately compound or simple. Leaves weakly clustered at apex of branches, the leaflets with asymmetric base, the petiole base thickened, to 2 cm diameter. Some species caducifolious. Infructescence terminal or borne from the trunk. Fruit a bivalved capsule, to 15 cm long, indurate, somewhat woody, generally smooth, brown when mature. Seeds winged, numerous per fruit. The wood used to manufacture paper pulp, cardboard, banisters, cabinets, furniture, and particle board. Cultivated as an ornamental tree. The wood and bark contain high tannin content. Leaves, fruits, seeds, and bark medicinal. Distribution: Mexico and Belize to Peru and Bolivia.

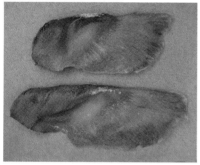

Bignoniaceae - *Clytostoma* (0.9x; 67.76 × 24)

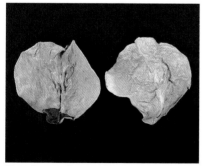

Bignoniaceae - *Clytostoma* (1.76x; 17.11 × 16.51)

Distictella O. Kuntze. Lianas. Leaves compound, opposite, trifoliate or bifoliate, with the third leaflet modified into a 3-branched tendril or in the form of a disk. Infructescence axillary or terminal. Fruit a bivalved capsule, to 20 cm long, typically smaller, weakly curved, smooth, the exocarp to 0.4 cm thick, tomentose or with very short, dense pubescence, with the texture of a chamois cloth, brown when mature. Seeds winged, numerous per fruit, in two rows. Plants with glands. Distribution: Costa Rica and Guianas to Peru and Argentina.

Bignoniaceae - *Jacaranda* (1.95x; 28.1 × 19.78)

Mussatia Bureau ex Baillon. Lianas. Stems square in some species. Leaves compound, opposite, trifoliate or bifoliate, the third leaflet modified into a simple tendril. Plants lacking glands. Pseudostipules very large and conspicuous. Infructescence terminal. Seeds winged, numerous per fruit, arranged in two rows within each of the valves of the capsular fruit. Distribution: Guianas, Colombia, and Brazil to Peru.

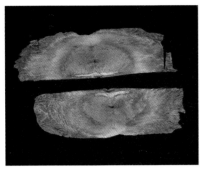

Bignoniaceae - *Mussatia* (1.4x; 40.23 × 12.15)

Paragonia Bur. Lianas. Leaves compound, opposite, bifoliate. Pseudostipules subulate. Tendrils two or 3-branched. Glands present on the petioles. Infructescence terminal or axillary. Fruit a bivalved capsule, to 50 cm long, subwoody, brown when mature. Seeds winged, many per fruit. Distribution: Mexico to Peru and Bolivia.

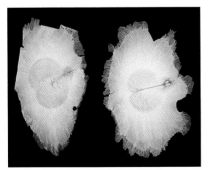

Bignoniaceae - *Paragonia* (0.9x; 51.5 × 34.49)

Periarrabidaea Samp. Lianas. Stems square in some species. Leaves compound, opposite, trifoliate or, if biofoliate, the third leaflet modified into a 3-branched tendril. Pseudostipules subulate. Glands present between the petioles. Fruit a bivalved capsule, thin, subwoody, brown when mature. Seeds winged, many per fruit. Distribution: Lowland Amazonian to Suriname.

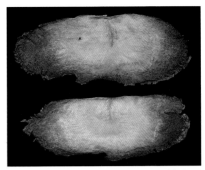

Bignoniaceae - *Periarrabidaea* (0.69x; 87.51 × 34.64)

Pithecoctenium Mart. Ex Meissn. Lianas. Stems square. Leaves compound, opposite, trifoliate or bifoliate with the third leaflet modified into a tendril with three or more branches. Infructescence terminal. Fruit a bivalved capsule, to 30 cm long, yellow to brown when mature, with indurate, conical bristles to 0.6 cm long. Seeds winged, numerous per fruit, in two rows. Plant with simple pubescence throughout or scattered. Distribution: Mexico and the West Indies to Paraguay and Argentina.

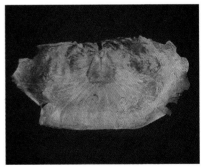

Bignoniaceae - *Pithecoctenium* (0.87x; 70.49 × 34.15)

Pleonotoma Miers. Lianas. Leaves compound, opposite, bifoliate, 2- to 3-ternate or 2- to 3-pinnate. Stems and branches conspicuously quadrangular. Pseudostipules present, leaving a circular scar after

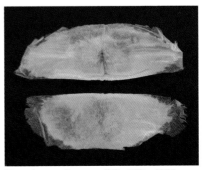

Bignoniaceae - *Pleonotoma* (0.96x; 57.91 × 19.34)

falling. Tendrils 3-branched. Infructescence axillary or terminal. Fruit a bivalved capsule, to 45 cm long, woody, brown when mature. Seeds winged, numerous per fruit. Leaves medicinal. Distribution: Venezuela, Colombia and the central Amazon to Peru.

Tabebuia Gomes ex DC. Shrubs or trees to 40 m tall. Leaves palmately compound, opposite, 5-foliate, rarely trifoliate or 7-foliate and very rarely with simple leaves. Some species caducifolious. Infructescence terminal or axillary. Fruit a bivalved capsule, to 30 cm long, brown when mature. Seeds winged, many per fruit. Some species shrub-like in savannahs and at the border of lakes with roots in water. Wood used for furniture and construction. Widely cultivated as ornamental trees and shrubs. Distribution: Mexico and Guianas to Peru and Bolivia.

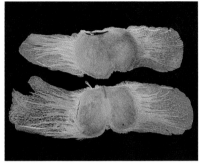

Bignoniaceae - *Tabebuia* (1.97x; 31.83 × 9.25)

Tanaecium SW. Lianas. Leaves compound, opposite, trifoliate or bifoliate with the third leaflet modified into a simple tendril. Infructescence terminal. Fruit a bivalved capsule, to 12 cm long, brown when mature. Seeds numerous per fruit, nearly lacking wings. Generally with a strong odor reminiscent of fish. Reportedly used as a hallucinogen. Distribution: Surinam and Guianas to Peru.

Bignoniaceae - *Tanaecium* (0.89x; 70.24 × 23.33)

BIXACEAE

Family of one genus of shrubs and trees. The inner bark oxidizes yellow-orange, and stems exude a residue of red or yellow sap. Leaves simple, alternate, with cordate base, the petiole with a swollen pulvinulus. Stipules present. Infructescence terminal. Fruit a bivalved loculicidal capsule, spinose or with bristles. Numerous red arillate seeds per fruit.

Bixa L. Shrubs or trees to 25 m tall. Leaves weakly 5-veined from base. Fruit to 6 cm long, brown or reddish when mature, covered with dense bristles to 1.2 cm long. Seeds numerous per fruit, covered with red-orange aril. Species common in secondary forests. Wood is light, used in production of cardboard and particle board. Amerindians use the red-orange aril for body paint, to color clothing, and to repel insects. The aril is also used widely to give a reddish color to food. A glue similar to gum arabic is obtained from the stems. Aril and seeds medicinal. Distribution: Native to tropical America, from Mexico to Paraguay, with one species widely cultivated across the tropics and subtropics of the world, to 1880 m elevation.

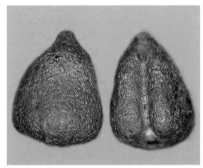

Bixaceae - *Bixa* (7.67x; 5.12 × 3.56)

BOMBACACEAE

Trees. Leaves alternate, simple, or palmately compound, with caducous stipules. Fruit a loculicidal capsule with five valves, woody, with many seeds surrounded by a cotton-like structure; or winged in *Huberodendron*, or a berry with five seeds embedded in a fleshy, fibrous mesocarp or samaroid in *Cavanillesia*. Some genera with spines on the trunk or stems. Some genera caducifolious. Many genera appreciated for their wood, the fibers of the fruit, or as ornamentals.

Cavanillesia Ruiz and Pavon. Trees to 30 m tall caducifolious. Trunk smooth with prominent circular scars, cylindrical, nearly without buttresses, and with a conspicuous swelling near the base giving an echo-forming metallic sound when hit; the bark orange-green, with a thin papyraceous, defoliating exterior that flakes away easily when rubbed. Leaves simple, coriaceous, the base cordate, clustered at the extreme apices of the branches. Fruit a samara to 14 cm long, the five wings membranous, brown when mature, tomentose. The seeds are edible and with pressure one can extract an oil that is used as a medicine and in cooking. Distribution: Nicaragua to Peru and Bolivia.

Bombacaceae - *Cavanillesia* (0.45x; 130 × 120)

Ceiba Mill. Trees to 60 m tall. Trunks to 3–4 m diameter. Leaves palmately compound, with 3–9 leaflets, clustered at the apices of the branches. Fruit a woody, 5-valved capsule, to 25 cm long, brown when mature. Seeds many per fruit, surrounded by white or brown cotton-like fibers. The trunk buttresses very large, sometimes up to 15–20 m tall. The majority of species with prickles on the branches and trunk, especially when juveniles. Some species caducifolious, others with swollen trunks. Occurring in seasonally inundated forests along floodplains in old clearings, and in dry forests. The wood is used to construct canoes, and is commonly used in the boxing and packaging industry. Some people eat the seeds roasted or ground, and oil from the seed is used to make soap. The fibrous hairs from within the capsular fruit has been used widely in life vests, sleeping bags, pillows, and mattresses. Roots medicinal. Distribution: Mexico and Cuba to Peru, also western Africa and in the Malesian Peninsula, to 2600 m elevation.

Bombacaceae - *Ceiba* (1.35x; 11.41 × 9.84)

Chorisia H. K. B. Trees to 40 m tall. Trunk with pronounced buttresses and swollen base. Leaves palmately compound with 5–7 leaflets, the margins entire or serrate. Fruit a woody, trivalved capsule, to 25 cm long, brown when mature. Seeds many per fruit, surrounded by white cotton-like fibers. The majority of species have prickles on the stems and trunk, especially in the juvenile stage, and are caducifolious. Found in forests that are seasonally inundated or not. Distribution: Tropical America to 1800 m elevation.

Bombacaceae - *Chorisia* (1.95x; 7.5 × 7.23)

Eriotheca Schott and Endl. Trees to 30 m tall, caducifolious. Trunk with small to well-developed buttress roots, and some prickles scattered throughout. Leaves palmately compound, with 5–7 leaflets, clustered in the apices of the branches, light pubescent with stellate hairs. Fruit a woody, 5-valved capsule, to 7 cm long, brown when mature. Seeds many per fruit, surrounded by a mass of brown fibers. Distribution: Brazil, Colombia, to Peru and Bolivia.

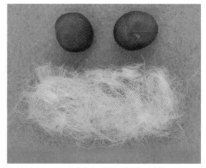

Bombacaceae - *Eriotheca* (1.69x; 9.25 × 7.88)

Huberodendron Ducke. Trees to 45 m tall. Leaves simple, entire. Stipules caducous. Fruit a 5-valved woody capsule, to 20 cm long, brown when mature. Seeds many per fruit, with large wings. Distribution: Central America and Colombia to Peru and Bolivia.

Bombacaceae - *Huberodendron* (0.8×; 76.42 × 23.05)

Matisia Bonpl. Trees to 35 m tall. Leaves simple, ovate to cordate. Fruit a 5-valved capsule, in some species a berry to 8 cm diameter, brown, yellow, or orange when mature. Seeds up to five per fruit. Some species with a persistent calyx, very well-developed, forming a cupule subtending the fruit, sometimes covering up to half of the fruit. Some species with well-developed buttress roots. Some authors include this genus in *Quararibea*. Distribution: Guianas to Peru.

Bombacaceae - *Matisia* (2.07×; 29.49 × 15.92)

Bombacaceae - *Matisia* (2.07×; 26.79 × 10.66)

Ochroma Sw. Trees to 20 m tall. Leaves simple and weakly lobed, the base cordate. Pubescence stellate or dendritic and very dense on the abaxial side of the leaves. Fruit a 5-valved capsule, to 25 cm long, brown when mature. Seeds many per fruit surrounded by brown cotton-like fibers. Stipules persistent. Trees common in secondary forests and along the banks

and floodplains of rivers. Species are exploited for their light wood in building model airplanes, flotation devices, toys, and boxes; the fibrous hairs from the fruits are used for stuffing pillows and mattresses; the strong fibrous bark is used as rope; and the leaves are a source of tannins. Distribution: Mexico and the Antilles to Peru and Bolivia. Cultivated widely throughout the tropics.

Bombacaceae - *Ochroma* (3.54×; 4.75 × 2.3)

Pachira Aubl. Trees to 30 m tall. Leaves palmately compound with 5–8 leaflets, clustered at the apices of branches. Fruit a woody 5-valved capsule, to 25 cm long, brown when mature. Seeds many per fruit, without wings or cotton-like fibers. Some species found commonly at the borders of lakes and other wetlands. Some species caducifolious. Seeds are sometimes eaten fried or baked. The wood is used to make decorative handicrafts, furniture, cabinetry, and in interior design and remodeling. Distribution: Mexico to Peru and Bolivia.

Bombacaceae - *Pachira* (0.96×; 36.59 × 23.08)

Bombacaceae - *Pachira* (1.32×; 26.88 × 21.6)

Pseudobombax Dugand. Trees to 25 m tall, caducifolious. Trunk whitish-gray with longitudinal green lines, and a characteristic swelling at the base. Leaves palmately compound with 5–7 leaflets, clustered at branch apices. Fruit a 5-valved capsule, to 25 cm long, brown when mature. Seeds many per fruit surrounded by brown cotton-like fibers. Distribution: Mexico to Bolivia and Peru, to 2200 m elevation.

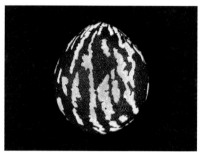

Bombacaceae - *Pseudobombax* (5.93x; 6.29 × 5.66)

Bombacaceae - *Pseudobombax* (1.57x; 6.28 × 5.47)

Quararibea Aubl. Trees to 25 m tall. Leaves simple, entire, trinerved, sometimes with a cordate base. Calyx persistent to 4 cm long covering half or more of the fruit. Infructescence borne from the stems. Fruit a berry, 3–5 cm diameter, yellow orange when mature. Seeds 1–2 per fruit. Branching in some species is characteristically verticillate. Distribution: Central America and Guianas to Peru and Bolivia.

Bombacaceae - *Quararibea* (2.6x; 17.68 × 10.65)

Bombacaceae - *Quararibea* (2.33x; 18.09 × 7.95)

BORAGINACEAE

Small family of shrubs, trees, herbs, and lianas. Leaves simple, alternate, rarely opposite or weakly clustered, sometimes pubescent and/or asperous to the touch. Some genera, such as *Cordia*, possess a solitary leaf arising at the point of apical bifurcation of the stems. Calyx persistent. Fruit variable, a drupe or dry berry with 2–4 locules. Seeds with indurate, solid endosperm.

Cordia L. Shrubs and trees to 25 m tall. Leaves simple, alternate and sometimes somewhat verticillate, entire or weakly serrate, glabrous or pubescent, some species asperous to the touch. Infructescence axillary or terminal. Fruit a drupe to 2 cm diameter, yellow or red to black when mature, the mesocarp fleshy, transparent. One species with drupaceous fruit with calyx and corolla persistent in the form of wings, brown when mature. Seeds one per fruit. Pubescence simple or compound. Some species with myrmecophilous swellings in the base of each ramification. In general a highly variable genus with much variation. The wood is used in construction and to make furniture. Leaves, seeds and bark medicinal. Distribution: Mexico and the Caribbean to Argentina to 2500 m elevation, also in the Old World.

Boraginaceae - *Cordia* (2.08x; 13.35 × 10.15)

Boraginaceae - *Cordia* (2.94x; 5.32 × 2.09)

Boraginaceae - *Cordia* (3.28x; 13.03 × 7.03)

Tournefortia L. Herbs or lianas. Leaves lightly asperous abaxially, opposite in some species. Infructescence terminal. Fruit sessile, to 1 cm diameter, white, with a strong aromatic odor. Seeds two per fruit. Leaves medicinal. Distribution: Honduras and Guianas to Peru, and the Old World.

Boraginaceae - *Tournefortia* (6.46x; 4.64 × 3.52)

BROMELIACEAE

Epiphytes, some species terrestrial. Leaves succulent, sessile, basal, alternate, spiral, many genera with spinose margins, lanceolate. Infructescence terminal, in some genera as large or larger than the plant. Fruit a

berry or capsule with tiny winged seeds. Many genera adapted to altitudes above 3500 m.

Aechmea R. & P. Epiphytic or terrestrial herbs. Leaves to 1 m long, green or red, margins with semi-erect spines. Infructescence 50 cm long, with basal spines to floral bracts red. Stigma persistent. Fruit a berry, sessile, yellow to bluish when mature. Seeds numerous per fruit. Distribution: Mexico to Argentina and Paraguay.

Ananas Mill. Terrestrial herbs. Leaves to 1 m long, the margin sometimes with spines. Infructescence to 50 cm long with spinose subtending bracts. Fruit a multiple formed by many fused simple fruits, yellow when mature. Seeds many per fruit. Plants widely cultivated and commercialized for their edible fruits. Distribution: Mexico to Argentina and Paraguay, with some species in Hawaii and the Old World tropics.

Bromeliaceae - *Ananas* (7.83x; 4.24 × 2.23)

Billbergia Thunb. Epiphytic herbs. Leaves to 1 m long, margins with spines. Infructescence to 1 m long. Stigma persistent. Fruit a berry, sessile, to 2.5 cm long, gray-white where mature, striate. Seeds numerous per fruit, embedded in a sticky mesocarp. Distribution: Costa Rica and Panama to Peru.

Bromeliaceae - *Billbergia* (7.2x; 4.59 × 2.68)

Guzmania R.& P. Epiphytic herbs. Leaves to 55 cm long, the margins entire, lacking spines, the venation fine and parallel without a conspicuous midvein, the petioles completely surrounding the stem and overlapping one another. Fruit a capsule, brown when mature. Seeds tiny, winged, many per fruit. Distribution: Southern United States to Peru and Bolivia.

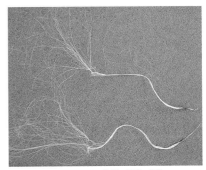

Bromeliaceae - *Tillandsia* (0.92x; 3.79 × 0.8)

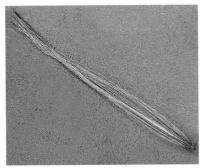

Bromeliaceae - *Guzmania* (4.44x; 17.48 × 1.26)

Neoregelia L. B. Sm. Epiphytic herbs. Leaves to 70 cm long, the apex acuminate, the margins spinose, the spines erect, black. Bracts spinose. Stigmas persistent. Calyx covering three-quarters the length of the fruit. Fruit a berry, white when mature. Seeds numerous per fruit, embedded in a sticky mesocarp. Distribution: Tropical America.

Bromeliaceae - *Neoregelia* (8.04x; 5.83 × 2.28)

Tillandsia L. Epiphytic herbs. Leaves to 25 cm long, the base wide and imbricate, borne from a swollen bulbous base, the margin lacking spines, green or mottled red. Infructescence with bracts. Fruit a capsule, to 6 cm long, brown when mature. Seeds tiny, winged, many per fruit. Distribution: Southern United States, Central America, Trinidad to Peru and Bolivia.

BURSERACEAE

Trees, rarely shrubs. Leaves alternate, pinnately compound, imparipinnate, the leaflets opposite, very pulvinulate or not. Fruit typically a capsule with 1–5 locules. Seeds surrounded by a white aril, or a drupe. Plant with resin, strongly aromatic, sometimes drying in the form of crystals in the trunk, sometimes with white latex. Some species develop aerial roots. Resin of some species used as a substitute for some petroleum products. Various species important for lumber.

Dacryodes Vahl. Trees to 30 m tall. The leaflets coriaceous with conspicuous pulvinuli and asymmetric bases. Infructescence terminal or axillary. Fruit a drupe to 2.5 cm long, yellow or brown when mature. Seeds one per fruit. Distribution: Tropical America, but more speciose in the Old World, to 1600 m elevation.

Burseraceae - *Dacryodes* (1.19x; 26.34 × 17.05)

Protium Burm. f. Shrubs or trees to 25 m tall. Leaves generally pinnately compound, sometimes trifoliate, one species unifoliate, the pulvinuli very conspicuous at both extremes of the petiolule. Infructescence terminal or axillary. Fruit a 2- to 4-valved capsule, to 2.5 cm long, red, yellow, or brown when mature, the interior of the valves pink. Seeds 1–3 per fruit with a white mealy aril. Some species with white latex and aerial roots. The wood of various species is used in construction, cabinetry, and woodwork. Distribution: Mexico, Central America, Guianas to Peru and Bolivia, to 1500 m elevation, pantropical.

Burseraceae - *Protium* (1.89×; 17.93 × 15.59)

Burseraceae - *Tetragastris* (3.72×; 13.92 × 8.78)

Trattinnickia Willd. Trees to 35 m tall. Leaves glabrous, or pubescent. Infructescence terminal. Fruit a drupe, to 1.5 cm diameter, black at maturity. Seeds one per fruit. Distribution: Panama and Guianas to Peru and Bolivia.

Burseraceae - *Protium* (1.82×; 19.71 × 13.26)

Burseraceae - *Trattinnickia* (1.97×; 9.46 × 9.14)

BUXACEAE

Small trees or shrubs. Leaves simple, alternate, or opposite, usually rather succulent or coriaceous. Infructescence axillary. Fruit a berry.

Styloceras Kunth ex. Juss. Trees to 16 m tall. Leaves alternate, entire. Fruit a berry to 3 cm diameter, yellow when mature, very distinctive by the two persistent, green, horn-like styles on each side of the fruit. Seeds various per fruit. Distribution: Colombia to Peru and Bolivia, to 3500 m elevation.

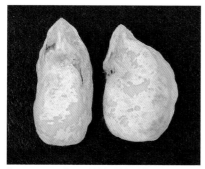

Burseraceae - *Protium* (4.04×; 10.6 × 6.2)

Burseraceae - *Protium* (2.2×; 8.99 × 5.1)

Tetragastris Gaertn. Trees to 25 m tall. Leaflets coriaceous, the bases asymmetric, short-petiolate. Infructescence terminal or axillary. Fruit a 1-valved or bivalved capsule, to 3 cm long, red when mature. Seeds 1–3 per fruit covered with a white mealy aril. Various species valued for their wood. Distribution: Central America and Guianas to Peru and Bolivia.

Buxaceae - *Styloceras* (2.86×; 9.63 × 6)

CACTACEAE

Terrestrial herbs and shrubs of dry areas, epiphytes in humid tropical forest, typically with succulent stems, spines, and often lacking true leaves, with transparent, mucose sap. Fruit a berry, usually red or white, with or without spines. Seeds numerous per fruit.

Epiphyllum Haw. Herbaceous epiphytes of forest canopy. Stems flattened with succulent sections, sometimes branching, the margin wide, irregularly undulate, principal vein prominent, secondary veins inconspicuous. Fruit a sessile, striate, edible berry, to 7 cm long, pink when mature, the mesocarp white. Seeds numerous per fruit. Distribution: Costa Rica to Paraguay.

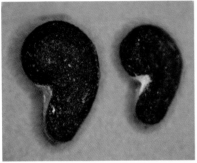

Cactaceae - *Epiphyllum* (13.54×; 3.02 × 1.82)

Selenicereus (A. Berger) Britton & Rose. Herbaceous epiphytes of forest canopy. Stems succulent, triangular in cross section, with groups of spines along two of the margins, the sections 80 cm long. Fruit a berry, spinose, to 12 cm long, yellow when mature. Seeds numerous per fruit. Cultivated for large, showy flowers. Distribution: Southern United States, Mexico, the Antilles, and Central America, to Peru and Bolivia.

Cactaceae - *Selenicereus* (14.29×; 3.2 × 1.65)

CAMPANULACEAE

Herbs. Leaves simple, alternate, serrate, with white latex. Fruit a capsule or berry. Majority of genera restricted to high altitudes, with one genus in lowland tropical forest. Well- represented in the Old World tropics.

Centropogon C. Presl. Succulent herbs to 3 m tall with hollow stems, sometimes pubescent throughout. Leaves weakly serrate. Infructescence axillary. Stigma persistent. Fruit a berry to 2 cm diameter. Seeds numerous per fruit. Reported to be cultivated for edible leaves. Distribution: Mexico, Central America, Antilles, to Chile, normally growing in pastures and other disturbed secondary growth areas in the Amazon; one or more species of wetlands.

CANNACEAE

Herbs. Leaves simple, alternate, the base of the petiole sheathing the stem. Infructescence terminal. Fruit a loculicidal capsule.

Canna L. Herbs to 2 m tall. Leaves to 1 m long. Fruit a trivalved capsule, to 3.5 cm diameter, brown when mature. Seeds 1–2 per locule. Plants with rhizomes. Widely cultivated as ornamentals. Distribution: Tropical America.

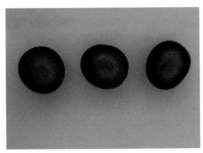

Cactaceae - *Canna* (2.39×; 6.77 × 6.99)

CAPPARACEAE

Herbs, shrubs, small trees, rarely lianas. Leaves alternate, simple or compound, 3- to 5-foliate, margins entire. Leaves and petioles variable in size on the same stem. Stipules present or absent. Infructescence axillary or terminal. Fruit a berry or capsule, stipitate. Some genera with small glands on the base of the petioles, others with spines. Glabrous or pubescent, the hairs simple or stellate. The name Capparidaceae used historically by some authors.

Capparis L. Shrubs or trees to 6 m tall. Leaves simple, the margins entire. Infructescence axillary or terminal. Fruit a berry, to 6 cm long, yellow when mature. Mesocarp transparent white. In some species the fruit is dark purple and opens irregularly at maturity exposing the white mealy mesocarp, from which hang the seeds. Seeds 1–16 per fruit. Some species considered important as forage, ornamentals, and for reforestation. Leaves, fruits, roots, and bark medicinal. Distribution: Brazil, Bolivia, and Peru to 2000 m elevation. Well-represented in the Old World. Common in dry forest.

Capparidaceae - *Capparis* (2.37x; 11.38 × 9.42)

Capparidaceae - *Capparis* (6.26x; 8.48 × 5.12)

Cleome L. Herbs to 2 m tall. Leaves simple or compound with 3–7 leaflets. Infructescence axillary or terminal. Fruit a capsule, to 4 cm long, yellow, open-

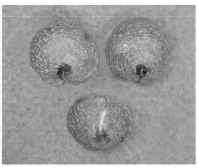

Capparidaceae - *Cleome* (11.11x; 2.15 × 2)

ing at maturity. Seeds 6–30 per fruit, clustered with no particular order, falling from the dehiscing fruit at maturity, leaving a complete margin. Some species with spines or glands. Distribution: Mexico and Surinam to Paraguay, also represented in the Old World to 3500 m elevation.

Crataeva L. Shrubs or small trees to 5 m tall. Leaves compound, trifoliate, the leaflets somewhat succulent. Fruit a berry. Seeds various per fruit. Plants lightly aromatic. Common in inundated or swamp forests. Distribution: Tropical America but more common in the Old World.

Capparidaceae - *Crataeva* (1.93x; 10.71 × 6.59)

Morisonia L. Shrubs to 6 m tall. Leaves simple, margins entire. Infructescence borne from the trunk. Fruit a berry, to 5 cm diameter, gray-green when mature, more woody than in *Capparis*, the mesocarp white. Seeds 1–3 in small fruits and up to 20 in large fruits. Distribution: Brazil, Bolivia, and Peru.

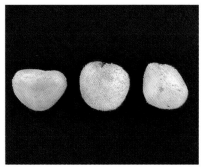

Capparidaceae - *Morisonia* (1.44x; 13.09 × 11.4)

CARICACEAE

Shrubs and trees, rarely lianas. Trunk composed of weak, spongy, soft wood, sometimes with prickles, very rarely lianas. Leaves alternate, simple, multilobed, or palmately compound, sometimes with abundant latex. Fruit a fleshy berry, yellow or orange when mature, with many seeds.

Carica L. Shrubs or trees to 10 m tall; one species a liana. Trunk not branched. Leaves simple, 3- to 5-lobed, variable in size and form, some species caducifolious. Latex white to aqueous. Infructescence axillary or borne from the trunk. Fruit a berry, to 40 cm long in cultivated species, smooth or somewhat five-angled, green, yellow, red, or orange when mature. Seeds numerous per fruit. Distribution: Panama to Brazil, and Argentina, to 3000 m elevation.

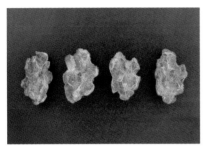

Caricaceae - *Carica* (2.31×; 7.84 × 5.32)

Jacaratia A. DC. Trees to 25 m tall. Trunk cylindrical with spongy bark and large conical prickles. Leaves sometimes simple, lobed, usually palmately compound with 4–7 leaflets, the petiolar base and stems with conical prickles. Infructescence axillary. Fruit a berry, to 18 cm long, orange when mature, with abundant white latex and orange mesocarp. Seeds numerous per fruit, each surrounded by a sticky, transparent aril. Common in forest gaps and areas of secondary growth. Distribution: Costa Rica to Peru and Bolivia.

Caricaceae - *Jacaratia* (5.82×; 6.8 × 4.13)

CARYOCARACEAE

Large trees. Leaves compound, trifoliate, alternate or opposite, glabrous or pubescent. Fruit a drupe. The wood of this family is very strong and appreciated in the construction of boats and handles of various types of hand tools.

Anthodiscus G. F. Meyer. Trees to 35 m tall. Leaves alternate, the margins sometimes crenate, the stipules leaving an interpetiolar scar. Infructescences ter-minal. Fruit a drupe to 2 cm diameter, green-yellow when mature, 8–15 locular, with woody pericarp. Seeds one per fruit. Wood used as parquet, posts, and beds. Distribution: Brazil, Colombia, Peru, and Venezuela.

Caryocaraceae - *Anthodiscus* (1.26×; 26.47 × 25.46)

Caryocar L. Trees to 30 m tall. Leaves opposite, the margins sometimes serrate, sometimes with petiolar glands, the stipules leaving a circular scar. Fruit a drupe to 8 cm long, yellow-green when mature, the mesocarp yellow, mealy, the endocarp woody, tuberculate, or spinose. Seeds 1–2 per fruit. Commercialized for wood and edible fruits of some species. Seeds and bark medicinal. Distribution: Costa Rica to Bolivia and Paraguay.

Caryocaraceae - *Caryocar* (0.88×; 50 × 33.37)

Caryocaraceae - *Caryocar* (0.71×; 52.3 × 45.48)

CECROPIACEAE

Trees and hemiepiphytes. Leaves simple, alternate, entire, or lobed. Pubescent or glabrous. Stipule very conspicuous, encircling stem. Latex or transparent sap oxidizing dark brown. Infructescence terminal. Fruit a drupe or achene in a fleshy structure. Mostly composed of heliophytic species, pioneers of disturbed forest. Some genera with aerial roots.

Cecropia Loefl. Trees to 30 m tall. Leaves peltate, 5- to 9-lobed; one species with 12–18 lobes completely separated, with petiolules. The terminal stipule conspicuous. Infructescence with a few to many spikes, fleshy, to 20 cm long, green or yellow when mature. Fruit an achene. Seeds one per fruit. Stems with conspicuous rings. Many species with aerial roots to 1 m long. Glabrous or pubescent to almost spinose on the leaves and stems. The genus consists of pioneer species of secondary forest growth, natural clearings, or the borders of rivers and lakes. Many species are associated with ants, which feed on structures that exude sugars in the base of the petioles. Various species have wood that is useful for producing paper pulp. Young leaves, shoots, and bark medicinal. Distribution: Mexico and Surinam to Peru and Bolivia, to 2000 m elevation.

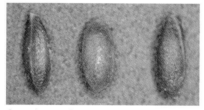

Cecropiaceae - *Cecropia* (16.19x; 1.69 × 0.72)

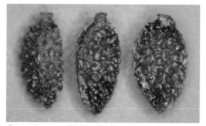

Cecropiaceae - *Cecropia* (13.97x; 2.42 × 1.13)

Cecropiaceae - *Cecropia* (9.05x; 4.68 × 1.54)

Coussapoa Aubl. Hemiepiphytes, rarely trees. Leaves simple, normally glabrous, sometimes scabrous or pubescent, margins entire; the tertiary venation very fine and parallel, perpendicular with the secondary veins. Terminal stipules caducous, leaving circular scars on the stem. Infructescence globose heads, axillary, in racemes or solitary, to 3 cm diameter. Fruit an achene, yellow, red, or green outside and red inside at maturity. Seeds one per fruit. Distribution: Central America and Guianas to Bolivia and Peru, to 2400 m elevation.

Cecropiaceae - *Coussapoa* (10.37x; 2.89 × 1.64)

Cecropiaceae - *Coussapoa* (7.78x; 4.45 × 2.01)

Pourouma Aubl. Trees to 25 m tall. Leaves simple, entire, or with up to 11 lobes, glabrous or sometimes pubescent or scabrous, the tertiary venation very fine and parallel, perpendicular to the secondary veins. Terminal stipule caducous, leaving circular scars on

Cecropiaceae - *Pourouma* (3.32x; 12.17 × 7.1)

the stem. Infructescence axillary, racemose. Fruit a drupe, to 2 cm diameter, glabrous, asperous, or pubescent, yellow, dark purple, or black when mature. Seeds one per fruit. Many species with aerial roots to 1 m. Genus of pioneer species, common in secondary forest. Some species cultivated for their edible fruits. Distribution: Nicaragua and Guianas to Bolivia and Peru, to 1800 m elevation.

Cecropiaceae - *Pourouma* (2.54×; 14.46 × 9)

CELASTRACEAE

Shrubs and small trees, one genus of lianas. Leaves alternate and simple, opposite in one genus in Central America. Terminal stipule caducous. Infructescence terminal or axillary. Fruit a bi- trivalved capsule with 1-3 seeds surrounded by a white, red, or orange aril, or a samara.

Gymnosporia (Wight & Arn.) Hook f. Trees to 10 m tall. Leaves alternate, entire. Infructescence axillary or sometimes borne from the trunk. Fruit a trivalved capsule, to 1 cm diameter, yellow when mature. Seeds 1–3 per fruit, surrounded by a white aril. Segregated from the genus *Maytenus* . Distribution: Peru, better represented in the Old World tropics.

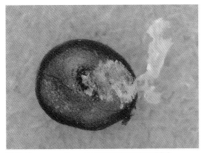

Celastraceae - *Gymnosporia* (5.71×; 6.22 × 5.26)

Maytenus Molina. Trees to 25 m tall. Leaves alternate, entire. Infructescence axillary. Fruit a bi- to trivalved capsule, to 2 cm diameter, yellow or red when mature. Seeds 1–2 per fruit, surrounded by a white or red aril. Some species with spines. Distribution: Trop-

ical and subtropical America, also in Asia and Africa, to 3800 m elevation.

Celastraceae - *Maytenus* (3.41×; 13.23 × 10.43)

CHLORANTHACEAE

Aromatic trees. Leaves simple, opposite, the margin weakly to strongly serrate, the petiole bases fused around the stem giving the impression of an ocrea and interpetiolar swellings. Stipules persistent. Infructescence axillary or terminal. Fruit a berry.

Hedyosmum Sw. Shrubs or trees to 8 m tall. Leaves lightly carnose or succulent, opposite and decussate, exuding lemon fragrance when crushed. Fruit a berry to 2 cm long, fleshy, yellow to white when mature. Seeds small, many per fruit. A genus predominantly of cloud forests, however at least one species forms an important component of palm swamp wetlands in Madre de Dios, Peru, below 300 m elevation. Distribution: Mexico to Peru and Bolivia, also in Asia, to 3500 m elevation.

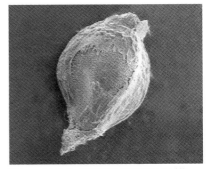

Chloranthaceae - *Hedyosmum* (17.04×; 3.1 × 1.92)

CHRYSOBALANACEAE

Shrubs and trees. Sometimes with white or reddish latex. Inner bark with a sandy texture. Stipules sometimes inconspicuous, persistent or caducous. Leaves simple, alternate, entire. Glands often present in pairs on the petiole or at the base of the lamina, rarely on the lamina. Glabrous or pubescent. Infructescence axillary or terminal. Fruit a drupe, dry or fleshy, the endocarp woody or thin. Seeds one per fruit.

Couepia Aubl. Shrubs or trees to 20 m tall. Leaves often lightly pubescent abaxially, glabrous adaxially. Stipules axillary, very conspicuous, sometimes caducous. Some species with exudate. Infructescence axillary or terminal. Fruit to 10 cm diameter, brown when mature, dry, or fleshy. Glands sometimes present on the petioles. Some species with edible fruits. Distribution: Mexico and Guianas to Peru and Bolivia.

Chrysobalanaceae - *Couepia* (1.1x; 50.19 × 23.94)

Hirtella L. Shrubs or trees to 30 m tall. Leaves always with reticulate venation and abaxially pubescent or glabrous. Stipules caducous or persistent. Some species with red exudate. Infructescence terminal, sometimes axillary. Calyx and stamens persistent. Fruit to 3 cm long, yellow or dark purple when mature, the mesocarp fleshy, juicy. Some species myrmecophilous. Some species valued for their wood. Distribution: Mexico and Western Indies to Peru and Bolivia, also in Africa and Madagascar.

Chrysobalanaceae - *Hirtella* (2.57x; 13.99 × 7.57)

Chrysobalanaceae - *Hirtella* (1.88x; 15.28 × 8.37)

Licania Aubl. Trees to 35 m tall. Sometimes with aerial roots. Rarely with white latex, sap reddish or transparent. Leaves usually coriaceous. Stipules axillary, sometimes caducous. Infructescence axillary or terminal. Fruit to 8 cm diameter, green or brown when mature, the mesocarp white, yellow, or orange. Plants glabrous or short pubescent, sometimes with glands on the petioles. Wood generally very strong and resistant, used in marine construction, to make railroad ties, and in the production of charcoal. Some species with edible fruits. Distribution: Panama and Guianas to Peru and Bolivia, one species in Asia.

Chrysobalanaceae - *Licania* (1.09x; 26.99 × 26.21)

Chrysobalanaceae - *Licania* (1.63x; 30.91 × 30.26)

CLUSIACEAE

Trees, shrubs, and hemiepiphytes, rarely lianas. Leaves simple, opposite, entire, one genus with alternate leaves. Latex yellow, orange, red, or white, sometimes transparent and resinous. Fruit a drupe, berry, or capsule. Seeds arillate or not, sometimes winged. Some genera with aerial roots. Secondary and tertiary leaf venation very fine and parallel, which is an important characteristic for recognizing many genera within this family.

Calophyllum L. Trees to 35 m tall. Leaves opposite. Latex yellow. Infructescence axillary or terminal. Fruit a drupe, to 3 cm diameter, green when mature, sometimes with a glaucous covering, the mesocarp white. Seeds one per fruit. Leaves with very fine and tightly parallel secondary venation, nearly perpendicular to the midvein. Wood very resistant to humidity and moisture, used in the construction of houses and canoes, columns, furniture, and floors. The latex and oil is used as a medicinal balm. Distribution: Mexico and West Indies to Peru and Bolivia, but most speciose in Asia and Micronesia.

Clusiaceae - *Calophyllum* (1.59×; 21.67 × 17.47)

Caraipa Aubl. Shrubs and trees to 30 m tall. Leaves alternate, rarely opposite, with translucent punctations, the tertiary venation parallel and perpendicular to the secondaries. Latex scant, resinous, and colorless. Infructescence axillary or terminal. Stamens persistent. Fruit an asymmetric septicidal capsule with three locules, to 2 cm long, brown when mature. Seeds winged, various per fruit. The wood important in construction of houses. Distribution: Brazil and Bolivia.

Clusiaceae - *Caraipa* (2.58×; 14.71 × 10.41)

Chrysochlamys Poeppig and Endl. Small trees to 8 m tall. Leaves opposite, succulent to coriaceous. Latex white or colorless. Infructescence terminal or borne from the trunk. Fruit a 4- to 5-valved capsule,

Clusiaceae - *Chrysochlamys* (4.13×; 7.51 × 3.74)

to 2 cm diameter, yellow-reddish when mature, the mesocarp white, milky. Seeds 2–4 per fruit, arillate or not. Distribution: Panama to Peru and Bolivia.

Garcinia L. Trees to 20 m tall. Leaves opposite. Infructescence borne from the stems. Latex yellow or white. Fruit a berry to 7 cm long, yellow or orange when mature, the exocarp smooth or warty, the mesocarp white, sweet. Seeds 1–4 per fruit. Leaves with reticulate venation. Some authors still maintain the genus (see *Rheedia*). Fruit edible, sold in local markets. Distribution: Panama and Guianas to Bolivia, but more speciose in the Old World tropics.

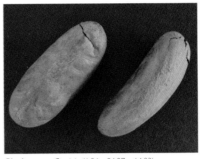

Clusiaceae - *Garcinia* (1.26×; 34.37 × 14.03)

Clusiaceae - *Garcinia* (1.76×; 22.53 × 12.53)

Havetiopsis Planch. et Triana. Hemiepiphytes, rarely shrubs. Leaves opposite, with fine, parallel secondary venation, almost imperceptible. Latex yellow. Infructescence terminal. Styles persistent. Fruit a 4-valved capsule, to 1.5 cm long. Seeds numerous per fruit with a red aril. Distribution: Panama to Bolivia and Peru.

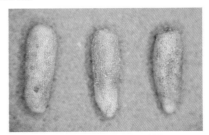

Clusiaceae - *Havetiopsis* (11.01×; 2.71 × 0.93)

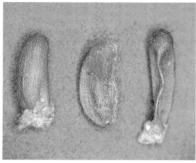

Clusiaceae - *Havetiopsis* (8.78x; 4.37 × 1.42)

Clusiaceae - *Rheedia* (2.07x; 22.45 × 10.97)

Marila Sw. Trees to 20 m tall. Leaves opposite, with the tertiary veins perpendicular to the secondaries, clearly anastomosing. Latex yellow, rapidly oxidizing brown. Infructescence axillary. Fruit a septicidal capsule, to 2 cm long, elongate, yellow or brown when mature. Seeds winged, numerous per fruit. Distribution: Guatemala to Bolivia and Peru.

Symphonia L. f. Trees to 35 m tall with aerial roots to 1 m long. Leaves opposite with fine, parallel secondary venation. Latex abundant, yellow. Stigma persistent. Fruit a berry to 4.5 cm diameter, green to reddish when mature, the exocarp with yellow latex. Seeds 1–3 per fruit. Wood used for firewood, carpentry, and train tracks. Latex is used to seal seams on boats. Latex medicinal. Distribution: Central and South America, West Indies, West Africa and Madagascar, to 1000 m elevation.

Clusiaceae - *Marila* (32.59x; 0.79 × 0.24)

Clusiaceae - *Symphonia* (1.78x; 23.93 × 17.2)

Tovomita Aubl. Trees to 20 m tall. Leaves opposite, somewhat succulent, with rather inconspicuous venation. Latex yellow or orange. Infructescence terminal or borne from the stems. Stigma persistent. Fruit a 4-valved septicidal capsule, to 7 cm diameter, green

Clusiaceae - *Marila* (34.92x; 0.79 × 0.24)

Rheedia L. Trees to 10 m tall. Leaves opposite, coriaceous, the secondary veins almost perpendicular to the midvein. Latex white or yellow. Infructescence axillary or borne from the stems. Stigma persistent. Fruit a berry to 3 cm diameter, yellow when mature, the exocarp smooth or warty, the mesocarp white, sweet. Seeds 1–5 per fruit. Some species segregated to *Garcinia*, a genus primarily of the Old World tropics. Distribution: Mexico and Puerto Rico to Peru and Bolivia, also in Madagascar.

Clusiaceae - *Tovomita* (2.65x; 21.64 × 8.24)

externally and red internally when mature. Seeds four per fruit, covered by an orange aril. Aerial roots sometimes present. Distribution: Costa Rica and Guianas to Bolivia and Peru, to 1300 m elevation.

Clusiaceae - *Tovomita* (2.46x; 21.38 × 8.41)

Clusiaceae - *Tovomita* (2.54x; 16.04 × 7.75)

Vismia Vand. Trees to 15 m tall. Leaves opposite, usually pubescent, often tomentose. Latex orange to reddish. Infructescence terminal. Fruit a berry to 1.5 cm diameter, subtended by the persistent calyx and crowned by persistent styles, yellow or orange when mature. Seeds numerous per fruit. Generally growing in disturbed forests. Distribution: Mexico, Guianas, and Trinidad to Peru and Bolivia, also in Africa, to 2800 m elevation.

Clusiaceae - *Vismia* (14.6x; 2.2 × 0.71)

Clusiaceae - *Vismia* (12.28x; 2.34 × 1.26)

COCHLOSPERMACEAE

Shrubs or medium-sized trees, with one genus of herbs. Leaves simple, alternate, palmately lobed or sometimes palmately compound. Fruit a capsule. Seeds many per fruit. Generally growing in disturbed forests or on upper river banks.

Cochlospermum H. B. K. Trees to 25 m tall, caducifolious. Plants pubescent or glabrous. Stipules caducous. Some species with red sap. Infructescence terminal. Fruit a 3- to 5-valved loculicidal capsule, to 7 cm diameter, brown when mature, usually erect, very visible or conspicuous at the extremities of the canopy; the exocarp woody. Seeds numerous per fruit, each with a line of hairs in one plane around the margin. Distribution: Tropical America, Africa, Asia, and Australia.

Cochlospermaceae - *Cochlospermum* (1.43x; 4.96 × 2.44)

COMBRETACEAE

Trees, shrubs, and lianas. Leaves simple, opposite, alternate, or verticillate, often clustered at the apices of the branches, always with entire margins. Sometimes with small stipules. Infructescence axillary or terminal. Fruit a drupe or samara, quite variable across the family and within genera.

Buchenavia Eichl. Shrubs or trees to 40 m tall. Leaves alternate, clustered at branch apices. Infructescence axillary. Fruit a drupe to 3 cm long, yellow or brown when mature, the mesocarp juicy. Seeds one per fruit. Glands usually present on the petiole apex near the base of the lamina. Distribution: Panama and Puerto Rico to Peru and Bolivia, to 1500 m elevation.

Combretaceae - *Buchenavia* (2.07×; 17.86 × 8.27)

Combretum Loefl. Usually lianas, rarely shrubs. Leaves opposite, rarely alternate. Infructescence terminal and axillary. Fruit a samara to 4 cm diameter, brown when mature, with four poorly or well-developed wings. Some species with spines on the stem. Distribution: Mexico and West Indies to Argentina. Better represented in the Old World.

Combretaceae - *Combretum* (0.94×; 51.12 × 50.34)

Combretaceae - *Combretum* (1.09×; 47.77 × 43.88)

Terminalia L. Trees to 35 m tall. Leaves alternate, clustered at the apices of the branches. Infructescence axillary. Fruit a samara to 5 cm long, brown-

yellow to tan when mature, with two well-developed and 2–3 small, rudimentary wings; appearing drupe-like in some species. Trunk with outer bark papyraceous, defoliating or peeling in plates in some species, and often with large buttress roots. The wood valued for construction, train tracks, and furniture. One species widely cultivated as an ornamental. Distribution: Mexico and Trinidad to Bolivia and Peru. Better represented in the Old World.

Combretaceae - *Terminalia* (1.4×; 33.79 × 17.95)

Combretaceae - *Terminalia* (2.07×; 14.34 × 7.4)

Thiloa Eichler. Lianas. Leaves opposite. Infructescence axillary. Fruit a samara to 4 cm diameter, brown when mature, with four wings that are usually small, or rarely well-developed. Distribution: Lowland Amazonia.

Combretaceae - *Thiloa* (2.05×; 15.21 × 11.08)

Combretaceae - *Thiloa* (1.97x; 23.28 × 8.82)

COMMELINACEAE

Herbs, erect or climbing, perennial or annual. Leaves simple, alternate or entire, the base surrounding the stem, succulent. Infructescence terminal or arising from the soil, with a sheathing bract. Fruit a loculicidal capsule. Interpetiolar nodes on the stem.

Commelina L. Herbs to 1 m tall or lianescent. Leaves with nearly imperceptible secondary venation, parallel to the midvein. Calyx persistent. Infructescence terminal. Fruit a bi- to trivalved capsule, to 0.7 cm long, white and shiny when mature, the exocarp fragile. Seeds 1–3 per fruit with orange aril. Distribution: Tropical and subtropical America, but more speciose in Africa and Madagascar.

Commelinaceae - *Commelina* (7.72x; 4.92 × 3.4)

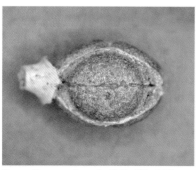

Commelinaceae - *Commelina* (8.68x; 5.94 × 3.33)

Commelinaceae - *Commelina* (8.57x; 5.96 × 3.42)

Dichorisandra J. C. Mikan. Herbs erect to 2 m tall or climbing. Leaves with inconspicuous secondary venation. Infructescence terminal. Calyx persistent. Fruit a trivalved capsule, to 2.5 cm long, purpleblack, yellow, or green when mature, the interior with purple valves. Seeds 3–6 per fruit, with orange aril. Pubescence short throughout. Leaves medicinal. Distribution: Mexico to Argentina.

Commelinaceae - *Dichorisandra* (6.72x; 5.34 × 3.17)

Floscopa Lour. Herbs to 0.5 m tall. Leaves with inconspicuous tertiary venation. Infructescence terminal. Fruit a capsule, to 0.5 cm diameter, shiny, discoid, white when mature, the shiny valves covered by bracts. Seeds one per locule. Distribution: Tropical America, but more speciose in the Old World.

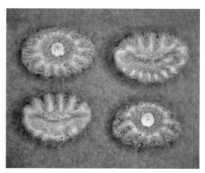

Commelinaceae - *Floscopa* (14.07x; 1.79 × 1.19)

CONNARACEAE

Lianas or lianescent shrubs, rarely small trees. Leaves alternate, pinnately compound, usually trifoliate, sometimes with 5–7 leaflets. Infructescence axillary or terminal. Fruit a follicle. Seeds one per fruit, with yellow or orange aril, rarely white. The leaflets with prominent, rounded pulvinuli similar to the Fabaceae. Plants glabrous or pubescent.

Connarus L. Lianas or shrubs. Leaves compound with 3–6 leaflets. Infructescence axillary or terminal, rarely borne from the trunk. Calyx persistent. Fruit a follicle to 2.5 cm long, red or yellow when mature, asymmetric with a mucronate apex. Aril covering one-third of the seed. Distribution: Costa Rica to Bolivia and Peru, also in the Old World.

Connaraceae - Connarus (3.56x; 10.6 × 6.85)

Pseudoconnarus Radlk. Lianas. Leaves compound with 1–3 leaflets. Infructescence axillary or borne from the stems. Fruit a follicle to 2 cm long, red when mature. Seeds one per fruit, covered one-third of its length by the aril. Distinct from the other genera by trinerved leaflets that are often glaucous abaxially. Distribution: Brazil, Bolivia, and Peru.

Connaraceae - Pseudoconnarus (4.19x; 13.21 × 5.86)

Rourea Aubl. Lianas or shrubs to 3 m tall. Leaves usually pinnately compound with 5–7 leaflets, sometimes trifoliate. Infructescence axillary or terminal. Calyx persistent covering the base of the fruit in the form of a cupule. Fruit a cylindrical erect or somewhat recurved follicle to 1.5 cm long, red when ma-

ture. Aril covering one-third of the seed. Distribution: Mexico and Cuba to Bolivia and Peru, also in the Old World.

Connaraceae - Rourea (5.66x; 10.36 × 4.43)

Connaraceae - Rourea (4.44x; 10.88 × 5.64)

CONVOLVULACEAE

Usually lianas, rarely herbs or shrubs. Leaves simple, alternate, rarely palmately compound. Some genera with stipules or white latex. Calyx persistent. Fruit a loculicidal capsule or a drupe. Possibly the only family that combines leaves with a cordate base, pinnately-nerved secondary venation, and relatively long petioles.

Calycobolus Willd. Ex. Schultes. Lianas. Leaves short pubescent. Stigma persistent. The calyx strongly lobed and of a papyraceous consistency

Convolvulaceae - Calycobolus (2.22x; 18.38 × 16.95)

covering the fruit. Fruit a drupe to 1 cm long, brown when mature. Distribution: Colombia, Brazil, Peru, and Bolivia, but more speciose in Africa.

Dicranostyles Benth. Lianas. Leaves simple. Infructescence axillary. Calyx persistent but not forming a cupule. Fruit a drupe to 3 cm long, black when mature, the exocarp subwoody, the mesocarp juicy. Plants glabrous or pubescent. Distribution: Guianas and Venezuela to Brazil, and Peru.

Ipomoea L. Lianas, rarely shrubs. Leaves simple with a cordate base. Latex white. Calyx persistent completely covering the fruit. Fruit a 4-valved capsule, to 1.5 cm long, brown when mature. Seeds four per fruit. Distribution: Mexico and West Indies to Argentina, also in the Old World tropics.

Convolvulaceae - *Merremia* (1.14x; 46.76 × 40.56)

COSTACEAE

Herbs, rhizomatous, to 5 m tall. Stems succulent with distinct internodes. Leaves alternate, spirally arranged along the stem axis, sessile or with short petioles sheathing the stem; the secondary venation parallel or inconspicuous. Plants pubescent or glabrous. Infructescence terminal with many fruits covered by large bracts of attractive colors and maturing over a rather long period of time. Fruit a loculicidal capsule. Seeds with an aril, various per fruit.

Costus L. Herbs to 4 m tall. Leaves usually in a perfect spiral around the stem, sometimes clustered at the stem apex, sessile or short-petiolate. Bracts red, green, or a combination of these colors, weakly superimposed, usually with a line at the apex. Calyx and stigma persistent. Fruit a trivalved capsule, to 2 cm long, white when mature. Seeds various per fruit, surrounded by a white, fibrous aril. Plants glabrous to densely pubescent. Distribution: Mexico to Peru and Bolivia, also in the Old World.

Convolvulaceae - *Ipomoea* (2.54x; 4.75 × 3.5)

Maripa Aubl. Lianas. Leaves simple. Infructescence axillary or terminal. Calyx persistent in the form of a cupule at the base of the fruit. Fruit an apiculate drupe to 2.5 cm long, black when mature, the exocarp subwoody, the mesocarp juicy. Seed one per fruit. Distribution: Panama and Guianas to Brazil, and Peru.

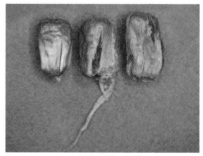

Costaceae - *Costus* (6.08x; 3.88 × 2.09)

Convolvulaceae - *Maripa* (2.92x; 13.15 × 9.35)

Merremia Dennst. ex Endl. Lianas. Leaves usually palmately compound, when simple the base cordate. Latex white. Calyx persistent covering one-third of the fruit. Fruit a 4-valved capsule, to 1 cm long, brown when mature. Seeds four per fruit. Pubescence white and short throughout. Distribution: Mexico to Bolivia and Peru, also in the Old World.

Costaceae - *Costus* (6.83x; 4.31 × 2.5)

Dimerocostus Kuntze. Herbs to 4 m tall. Leaves lanceolate, irregularly spiraled around the stem or clustered at the stem apex, the petiole base sheathing the stem leaving circular scars, the secondary venation fine, parallel. Infructescence terminal. Stigma persistent. Bracts superimposed but not as strongly clustered as in *Costus*. Fruit a bivalved capsule, to 3 cm long, white when mature. Seeds many per fruit, surrounded by white aril. Plants glabrous or glabrescent. Distribution: Honduras and Costa Rica to Peru and Bolivia.

Costaceae - *Dimerocostus* (9.52×; 4.31 × 2.49)

CUCURBITACEAE

Lianas or herbaceous climbers. Tendrils simple or branched. Leaves alternate, simple, entire, trilobed, or divided, sometimes trifoliately or palmately compound. Infructescence axillary. Fruit a special berry called a pepo, with a few too many seeds; in one genus a pixis with winged seeds.

Calycophysum Arn. Lianas. Tendril simple or branched. Leaves simple, cordate. Fruit a pepo to 11 cm long, orange when mature, the mesocarp orange. Seeds many per fruit. Distribution: Peru and Bolivia, to 2000 m elevation.

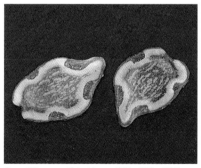

Cucurbitaceae - *Calycophysum* (3.64×; 8.92 × 5.93)

Cayaponia Silva Manso. Lianas. Tendrils branched, sometimes bifid or simple. Leaves simple, 3- to 7-lobed, rarely entire or trifoliately compound, the base cordate, the margins often dentate. Large bract at the base of the infructescence and many smaller bracts interspersed among the fruits. Stigma persistent. Fruit a pepo to 5 cm long, red or orange when mature, clustered in compact racemes, rarely solitary; the exocarp asperous, crustaceous, angled; the mesocarp white or yellow-orange, fibrous. Seeds 1–3 per fruit. Glands prominent, at the base of the leaf adaxially, sometimes more than two. Commonly the lamina is asperous to the touch. Distribution: Mexico and Surinam to all of tropical South America, also in Africa and Java.

Cucurbitaceae - *Cayaponia* (2.12×; 17.7 × 12.15)

Fevillea L. Lianas. Tendril bifid. Leaves simple or compound, trifoliate, margins often dentate; one pair of glands on the upper half of the petiole in species with compound leaves. Fruit a pepo to 10 cm diameter, green when mature, the mesocarp white. Calyx leaving prominent scars on the exocarp at the apex of the fruit. Seeds various per fruit. Plants pubescent throughout. Distribution: Costa Rica to Brazil and Peru.

Cucurbitaceae - *Fevillea* (0.43×; 52.07 × 49.59)

Gurania (Schltdl.) Cogn. Lianas. Tendrils simple. Leaves simple, lobed, or palmately compound with 3–5 leaflets; the margins often dentate, the lamina sometimes asperous to the touch. Stigma persistent.

Infructescence axillary or borne from the trunk. Fruit a pepo to 7.5 cm long, green-yellow with darker lines when mature, sometimes green and glaucous. Seeds many per fruit. Some species pubescent. Distribution: Central America to Bolivia and Peru.

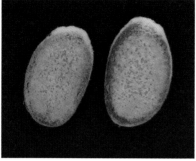

Cucurbitaceae - *Gurania* (5.04x; 7.31 × 5.69)

Melothria L. Lianas. Tendril simple. Leaves simple, entire or 3- to 5-lobed, the base cordate, the adaxial lamina surface asperous to the touch, the margins often weakly dentate, glands absent. Fruit a pepo to 3.5 cm diameter, yellow or orange with darker lines, sometimes black when mature. Seeds many per fruit embedded in an orange pulp. Distribution: Southern United States to Argentina.

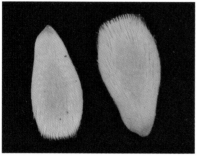

Cucurbitaceae - *Melothria* (5.12x; 7.87 × 4.13)

Siolmatra Baill. Lianas. Tendril bifid. Leaves compound with 3–6 leaflets, coriaceous; the petioles auriculate-glandular at or below the middle, to 3 cm long, auriculate near the apex. Fruit a pseudopixis,

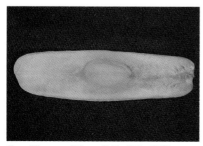

Cucurbitaceae - *Siolmatra* (1.11x; 55.9 × 15.33)

to 8 cm long, brown when mature. Seeds winged, many per fruit. Plants glabrous or pubescent. Distribution: Peru and Brazil.

CYCLANTHACEAE

Herbs and hemiepiphytes. Leaves simple, alternate or clustered, sometimes arising from the soil, petiolate, bifid or multi-lobed, rarely entire. Infructescence consisting of multiple fruits tightly aggregated and surrounded by one or various bracts. Fruit a berry. Seeds many per fruit. Petiole base sheathing the stem. Vegetatively easy to confuse with palms (Arecaceae).

Carludovica Ruiz and Pavon. Herbs acaulescent or the stem very short. Leaves 4-lobed, each pinnae parted in multiple segments, the petiole to 3 m long. Infructescence with peduncle arising from the soil, or from the short stem, to 25 cm long, erect; the spathe carnose, coriaceous, white, or reddish, opening and falling in pieces to expose the fruits. Fruit a berry to 1 cm long, orange or red when mature. Seeds many per fruit, covered by a red aril. The young leaves are used to obtain fibers for weaving hats. Distribution: Mexico to Peru and Bolivia.

Cyclanthaceae - *Carludovica* (11.43x; 2.02 × 1.03)

Thoracocarpus Harling. Hemiepiphytes. Leaves entire or bifid. Infructescence to 10 cm long, surrounded by many coriaceous bracts. Fruit a berry to 1 cm diameter, green or yellow when mature. Seeds many per fruit. Distribution: Brazil, Peru, and Bolivia.

Cyclanthaceae - *Thoracocarpus* (17.14x; 2.13 × 0.79)

CYPERACEAE

Herbs, usually erect, sometimes climbing. Leaves lanceolate, often with strongly serrate margins sharp and abrasive to the touch; the petiole generally elongate. Infructescence with bracts. Fruit a drupe. Differing from the grass family (Poaceae) by the triangular stems.

Diplasia Pers. Herb to 1.5 m tall. Leaves to 120 cm long, the margins with tiny transparent spines. Fruit sessile, to 1 cm long, brown when mature. Distribution: Guianas to Peru.

Cyperaceae - *Scleria* (5.08×; 4.6 × 3.08)

Cyperaceae - *Diplasia* (4.53×; 8.46 × 4.79)

Pleurostachys Brongn. Herbs to 1 m tall. Leaves with secondary venation parallel to the midvein, the margins with small spines, very fine along the midvein on the abaxial surface. Infructescence sometimes with spines and basal bracts. Fruit brown when mature. Seeds many per fruit, covered by various brown bracts. Distribution: Tropical South America.

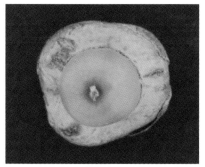

Cyperaceae - *Scleria* (5.02×; 7.68 × 6.98)

Cyperaceae - *Scleria* (4.95×; 7.88 × 8.1)

Cyperaceae - *Pleurostachys* (7.78×; 6.65 × 1.74)

Scleria Bergius. Herbs to 2 m tall, sometimes climbing. Plants with small spines throughout. Leaves with 3–5 conspicuous, finely parallel nerves; the margin abrasive, the petiole sheathing the stem. Infructescence axillary and terminal, with bracts. Fruit sessile, to 0.6 cm long, brown, black, or very shiny whitish-blue and smooth when mature, surrounded at the base by the persistent calyx. Distribution: Mexico and Cuba to Peru and Bolivia, also in Africa and Madagascar.

Cyperaceae - *Scleria* (5.08×; 4.96 × 3.35)

DIALYPETALANTHA-CEAE

Trees. Leaves simple, opposite, with two pairs of conspicuous stipules on each node. Fruit a loculicidal capsule. Caducifolious. Pubescent. Family very vegetatively similar to the Rubiaceae.

Dialypetalanthus Kuhlmann. Trees to 20 m tall. Leaves entire. Stipules present throughout, including the infrutescence. Infructescence terminal. Calyx lobes persistent. Fruit a 2- to 4-valved capsule, to 2 cm long, brown when mature. Seeds winged, many per fruit. Distribution: Brazil, Peru, and Bolivia.

Dialypetalanthaceae - *Dialypetalanthus* (2.12×; 17.4 × 7.75)

DICHAPETALACEAE

Shrubs, trees, and lianas. Leaves simple, alternate. Stipules persistent or caducous. Infructescence rarely axillary or terminal, usually on the petioles or at the base of the leaf blade, sometimes on the principal nerve on the lamina. Fruit a drupe. Pubescence of simple hairs often present.

Tapura Aubl. Shrubs or trees to 25 m tall. Leaves entire, more or less coriaceous. Calyx persistent. Fruit a drupe to 2.5 cm long, yellow when mature, arising from the petiole apex near the base of the lamina; the exocarp pubescent. Seeds one per fruit.

Dichapetalaceae - *Tapura* (2.86×; 16 × 8.59)

Plants glabrous or with short, yellow hairs throughout. Wood used in local construction. Majority of species have a toxic substance in the leaves and seed, sometimes used as a poison against rats and mice. Distribution: Mexico and Cuba to Bolivia and Peru, also in Africa.

DILLENIACEAE

Lianas, rarely shrubs. Stems usually orange-brown with a papyraceous bark, often asperous to the touch. Leaves simple, alternate, often asperous to the touch. Infructescence terminal, axillary, or borne from the stems. Fruit a follicle or capsule. Seeds arillate. Pubescence simple or stellate.

Doliocarpus Roland. Lianas. Pubescence simple. Leaves with reticulate venation, the petioles winged, the margins dentate. Infructescence axillary, or borne from the stems or trunk. Fruit a bivalved capsule, to 1.5 cm diameter, red when mature, subtended by persistent calyx. Seeds 1–2 per fruit, with white aril. Distribution: Mexico and Antilles to Paraguay.

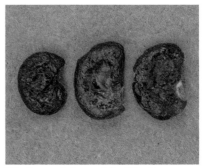

Dilleniaceae - *Doliocarpus* (2.65×; 8.56 × 6)

Tetracera L. Lianas. Bark defoliating. Pubescence stellate. Leaves asperous to the touch, the margins serrate, the petioles weakly winged. Infructescence axillary or terminal. Calyx persistent covering one-third of the fruit. Fruit a bivalved follicle, to 1 cm

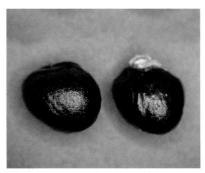

Dilleniaceae - *Tetracera* (7.09×; 3.84 × 3.76)

diameter, green or brown when mature. Seeds one per fruit with red, fibrous aril. Distribution: Mexico and Surinam to Paraguay, but more speciose in the Old World.

DIOSCOREACEAE

Herbaceous vines or lianas. Plants with rhizomes, sometimes well developed. Stems sometimes with spines. Leaves simple, alternate, entire or lobed, rarely completely dissected, the petioles elongate and sometimes twisted. Fruit a capsule, with three wings. Seeds winged, various per fruit.

Dioscorea L. Lianas. Leaves generally with a cordate base, the secondary venation arising from the base of the lamina. Infructescence axillary. Fruit a capsule, to 4 cm diameter, brown when mature, the locules in the form of wings, dehiscent. Seeds various per fruit. Widely cultivated for the edible tubers. Distribution: Tropical America, also in the Old World.

Dioscoriaceae - *Dioscorea* (2.31×; 1 2.57 × 12.52)

EBENACEAE

Trees. Leaves simple, alternate, often pubescent. Infructescence axillary. Calyx persistent and very conspicuous. Fruit a globose berry, glabrous or pubescent. Glands present on the abaxial surface of the leaves.

Diospyros L. Trees to 30 m tall. Leaves entire, the abaxial surface glaucous in some species. Infructescence axillary. Fruit a berry to 5 cm diameter, yellow, orange, or red when mature, surrounded at the base by persistent calyx with short, rounded lobes, the mesocarp white, sweet. Seeds 3–7 per fruit. Plants often pubescent. Various species coveted for the strong, black wood and edible fruit. Distribution: Panama and Trinidad to all of South America, but more speciose in Africa, Madagascar, and Asia.

Ebenaceae - *Diospyros* (2.86×; 19.52 × 9.63)

ELAEOCARPACEAE

Shrubs or generally trees with buttress roots that are generally large and slender. Leaves simple, alternate or often subopposite to opposite, clustered at branch apices, glabrous or pubescent, the trichomes simple or stellate; the petioles often elongate with two conspicuous swollen pulvini. Stipules persistent or caducous. Infructescence axillary. Fruit a capsule. Seeds arillate.

Sloanea L. Trees to 40 m tall. Leaves variable in size and shape. Stipules present in many species, caducous. Fruit a loculicidal capsule, 3- to 5-valved, rarely bivalved, to 4 cm diameter, or to 20 cm diameter in one species, yellow, green, red, or brown when mature, the exocarp smooth or with a sparse dense covering of soft to indurate bristles. Seeds usually one or rarely three per fruit, two-thirds of the length covered by an aril that is tan, orange, red, or green when mature. Many species with simple or stellate pubescence, short and dense on the leaf, petiole, and stems. Some species important for their wood. Buttress roots used to make oars or handles for tools. Distribution: Mexico and Antilles to Brazil, and Peru, also in Asia, to 1500 m elevation.

Elaeocarpaceae - *Sloanea* (1.31×; 22.74 × 22.98)

Elaeocarpaceae - *Sloanea* (1.7×; 17.14 × 15.65)

Elaeocarpaceae - *Sloanea* (1.97x; 22.2 × 11.13)

Erythroxylaceae - *Erythroxylum* (3.64x; 11.54 × 3.9)

Elaeocarpaceae - *Sloanea* (2.94x; 10.83 × 5.22)

ERYTHROXYLACEAE

Shrubs and trees. Leaves simple, alternate. Stipules conspicuous. Infructescence axillary or borne from the stems. Fruit a drupe. Seeds one per fruit. Usually easy to identify to family and genus by the presence of a band of often conspicuous veins running parallel to the midvein on the adaxial surface.

Erythroxylum P. Browne. Shrubs or trees to 18 m tall. Leaf margins entire with the central band of veins very conspicuous or inconspicuous. Calyx and stamens persistent. Fruit a drupe to 2 cm long, red when mature, the mesocarp fleshy, juicy, and red. Seeds one per fruit. Leaves of one species cultivated to obtain cocaine, prepare a tea, and chew. Distribution: Mexico and Cuba to Peru and Bolivia, also in the Old World, to 2000 m elevation.

EUPHORBIACEAE

Probably one of the most variable families in the Neotropics. Trees, shrubs, herbs, and lianas. One species with the aspect of Cactaceae. Leaves simple or trifoliately, compound, alternate, rarely opposite, often with glands, the margin entire or serrate, sometimes lobed; the petioles often elongate and variable in size. Stipules usually present, often caducous, sometimes absent. Often with latex or reddish to transparent sap, sometimes inconspicuous. Fruit a capsule, usually trivalved or with two, four, or more (up to 20) valves; some genera with drupes. Plants glabrous or pubescent; some genera with scales on the leaves, petioles, and stems. Family of major economic importance; used for wood, oils, resins, latex, and edible roots. Various genera important as ornamentals.

Acalypha L. Herbs or shrubs, sometimes lianas. Leaves alternate, the base cordate, or rounded, the margin crenate to serrate; often with 3–5 veins from the base; the petioles variable in size. Infructescence axillary or terminal with many bracts that sometimes cover the fruit. Fruit a trivalved capsule, to 0.5 cm diameter, brown when mature. Seeds various per fruit. Pubescence dense, sometimes white throughout. Plants generally growing in disturbed areas or secondary forest. Distribution: Tropical and subtropical America, also in the Old World, to 3000 m elevation.

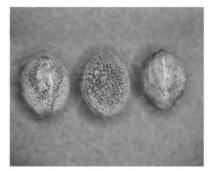

Euphorbiaceae - *Acalypha* (11.75x; 1.75 × 1.41)

Erythroxylaceae - *Erythroxylum* (2.84x; 16.42 × 6.79)

Adenophaedra (Muell. Arg.) Muell. Arg. Shrubs to 4 m tall. Leaves alternate, the margin serrate, the petioles usually elongate with an inconspicuous pulvinus at both extremes. Infructescence terminal. Fruit a trivalved capsule, to 1 cm diameter, brown when mature. Seeds 1–2 per fruit. Distribution: Amazonian lowlands.

Euphorbiaceae - *Adenophaedra* (3.26×; 9.09 × 7.99)

Alchornea Sw. Trees to 30 m tall. Leaves alternate with serrate margins, more or less 3-nerved from base, petiolate, with a pair of very obvious black glands at the base of the lamina on the abaxial surface. Infructescence axillary or borne from the stems. Fruit a bi- to trivalved capsule, to 0.9 cm diameter, green or brown when mature; the style persistent at apex. Seeds 2–3 per fruit, covered by red aril. Pubescence very short on the terminal branches, petiole, and veins, the trichomes simple or stellate. Distribution: Tropical and subtropical America, also in Africa, to 2500 m elevation.

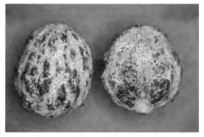

Euphorbiaceae - *Alchornea* (6.72×; 4.9 × 4.06)

Amanoa Aublet. Trees to 15 m tall. Leaves alternate, entire, coriaceous. Infructescence axillary. Fruit a trivalved capsule, woody, to 2.3 cm long, brown when mature. Seeds about four per fruit. Distribution: Tropical America, and the Old World.

Conceveiba Aubl. Trees to 15 m tall. Leaves alternate, the margin serrate, especially on the upper half of the lamina, the petiole usually elongate but quite variable, the pulvinus inconspicuous at both extremes; the lamina with glands at the base. Infructescence terminal. Fruit a trivalved capsule, to 2.5 cm diameter, green when mature, with three prominent lateral longitudinal wings, subtended by persistent calyx, terminated by three persistent stigmatic branches. Seeds 3–6 per fruit. Distribution: Guianas to Colombia, Ecuador, and Peru. One species in Africa.

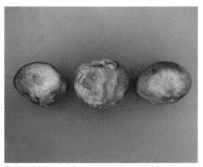

Euphorbiaceae - *Conceveiba* (1.86×; 10 × 8.81)

Croton L. Shrubs or trees to 20 m tall, rarely herbs. Leaves alternate with cordate or rounded base, petiolate or sessile, trinerved from the base in many species; usually but not always with a pair of conspicuous glands at the base of the lamina or the petiole apex. Resin red or absent. Infructescence axillary or terminal. Style persistent. Fruit a bi- to trivalved capsule, to 4 cm diameter. Seeds 2–3 per fruit, aril absent. Pubescence simple or stellate, very dense on the abaxial side of the leaf; the petiole and the branch apex sometimes glabrous but with white to tan scales on the abaxial surface of the lamina, very dense on the petiole and branch apex. Some species widely cultivated for the resin, which is used in medicine (*sangre de grado*). Distribution: Tropical and subtropical America, also the Old World, to 1800 m elevation.

Euphorbiaceae - *Croton* (1.65×; 32.63 × 24.23)

Euphorbiaceae - *Croton* (1.68×; 32.63 × 24.23)

Euphorbiaceae - *Croton* (6.61x; 3.22 × 2.64)

Drypetes Vahl. Trees to 30 m tall. Leaves alternate, entire, more or less coriaceous, the base generally asymmetric. Infructescence axillary. Fruit a drupe to 3.5 cm diameter, green or yellow when mature. Seeds one per fruit. Plants usually glabrous, rarely short pubescent. Distribution: Panama and Guianas to Peru and Bolivia, more speciose in the Old World.

Euphorbiaceae - *Drypetes* (1.49x; 26.67 × 17.23)

Euphorbiaceae - *Drypetes* (2.2x; 18.41 × 9.95)

Gavarretia Baill. Trees to 20 m tall. Leaves alternate, coriaceous, the margin often serrate on the upper half of the lamina; the petiole usually long but variable in size, the pulvinus inconspicuous at both extremes, the lamina with glands visible on the abaxial surface. Stipules caducous. Infructescence terminal. Fruit a bivalved capsule, to 2 cm diameter, green

when mature, with one prominent and longitudinal wing; the calyx and stigmas persistent. Seeds 1–2 per fruit. Distribution: Brazil and Amazonian Peru.

Euphorbiaceae - *Gavarretia* (3.03x; 9.93 × 8.5)

Glycidendron Ducke. Trees to 30 m tall. Leaves alternate, entire, coriaceous, weakly trinerved, with a pair of conspicuous, prominent concave glands on the adaxial surface. Stipules tiny. Latex in the young parts of the plant, new leaves, and new branches. Infructescence axillary. Fruit a drupe to 4 cm long, yellow when mature. Seeds one per fruit. Distribution: Surinam to Peru.

Euphorbiaceae - *Glycidendron* (1.28x; 40.99 × 17.52)

Hevea Aubl. Tree to 30 m tall. Leaves alternate, trifoliate compound, weakly grouped at the branch apices, with one pair of glands at the union of the three leaflets. Latex white, abundant. Stipules caducous.

Euphorbiaceae - *Hevea* (1.22x; 27.04 × 25.22)

Infructescence axillary. Fruit a trivalved capsule, to 5 cm diameter, brown when mature, woody, with explosive dispersal. Seeds 1–3 per fruit. Well known by the white resin used to make natural rubber. Distribution: Restricted to the Amazon but widely cultivated in all of the tropics of the world.

Hura L. Trees to 35 m tall. Trunk cylindrical, with many small prickles, especially when young or on the stems. Leaves alternate, cordate, petiolate, with a pair of conspicuous glands on the apex of the petiole at the junction of the lamina, the margins weakly serrate-undulate. Sap watery, yellow, and caustic. Stipule terminal, caducous. Infructescence axillary. Fruit a capsule with up to 15 valves, to 9 cm diameter, green when mature, woody. Seeds various per fruit. Wood used in carpentry, construction, plywood, and to make canoes or small boats. Leaves, seeds, resin, and bark medicinal. However, sap often said to be toxic, especially to the human eye. Distribution: Costa Rica and West Indies to Peru and Bolivia.

Euphorbiaceae - *Hura* (1.56x; 18.78 × 19.66)

Hyeronima Allemão. Trees to 30 m tall. Leaves alternate, petiolate, with prominent pulvinus near the base of the lamina and scales appearing like glandular punctations throughout the leaf and petiole, very dense on the abaxial surface. Stipules caducous, large on the young stems. Infructescence axillary. Fruit a drupe to 0.6 cm long, black and red when mature, the mesocarp fleshy, juicy. Seeds one per fruit. Distribution: Mexico to Peru and Bolivia, to 2500 m elevation.

Euphorbiaceae - *Hyeronima* (7.88x; 4.91 × 3.72)

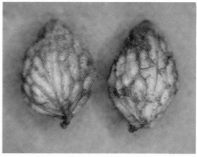

Euphorbiaceae - *Hyeronima* (7.67x; 4.87 × 3.48)

Mabea Aubl. Trees to 15 m tall. Branches rather verticillate. Leaves alternate, sometimes with glands, usually glaucous on abaxial surface, the margins entire or serrate, the venation anastomosing along the margins. Latex white, abundant. Stipules caducous. Infructescence terminal. Fruit a trivalved capsule, to 2.7 cm diameter, reddish or brown when mature, woody, the calyx and stigma persistent. Seeds three per fruit. Distribution: Mexico to Peru and Bolivia, to 1500 m elevation.

Euphorbiaceae - *Mabea* (1.57x; 12.09 × 11.22)

Manihot Mill. Shrubs or lianas. Leaves alternate, palmately 3- to 7-lobed, petiolate. Latex white or watery. Stipules small. Infructescence terminal or axillary. Fruit a trivalved capsule, to 2.5 cm diameter, green to brown when mature. Seeds 2–3 per fruit. Some species cultivated in Africa, Americas, and Asia for edible starchy roots (i.e., yucca). Distribution: Tropical America.

Euphorbiaceae - *Manihot* (2.86x; 11.74 × 9.19)

Maprounea Aubl. Trees to 25 m tall. Leaves alternate, the lamina relatively small, the petiole elongate; some species with glands on the base of the lamina. Stipules inconspicuous. Infructescence terminal. Fruit a trivalved capsule, to 0.5 cm diameter, brown when mature, each locule opening in two valves, the calyx persistent. Seeds two per fruit, weakly arillate. Distribution: South America, also in Africa, to 1500 m elevation.

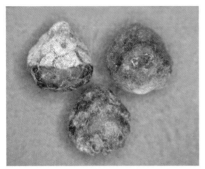

Euphorbiaceae - *Maprounea* (6.93x; 3.44 × 3.22)

Margaritaria L. f. Trees to 10 m tall. Leaves alternate, entire. Stipules small. Infructescence axillary. Fruit a 4- to 5-valved capsule, to 0.8 cm diameter, yellow or orange when mature, generally indehiscent, subtended by persistent calyx; the exocarp metallic blue and fragile; the mesocarp white. Seeds 5–8 per fruit. Distribution: Mexico and Guianas to Peru and Bolivia, also in the Old World.

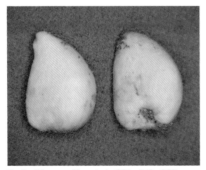

Euphorbiaceae - *Margaritaria* (9.21x; 3.68 × 2.59)

Nealchornea Huber. Trees to 28 m tall. Leaves alternate with glands at base of the lamina, the margin weakly serrate, the petioles quite variable in size. Infructescence axillary or subterminal. Fruit a berry, to 2.5 cm diameter, yellow when mature. Seeds 1-2 per fruit. Distribution: Brazil and Peru.

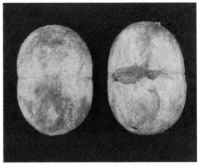

Euphorbiaceae - *Nealchornea* (2.6x; 14.59 × 10.2)

Omphalea L. Lianas of the forest canopy. Leaves alternate; with a pair of conspicuous glands at the junction of petiole and lamina on the adaxial surface; the base weakly cordate, the margin weakly undulate Sap abundant, watery, rapidly oxidizing. Infructescence axillary. Fruit a berry to 12 cm diameter, green when mature, the mesocarp white. Seeds 3–4 per fruit. Pubescence dense on the abaxial laminar surface, petiole, and branch apex. Distribution: Honduras and West Indies to Ecuador and Peru, also in the Old World.

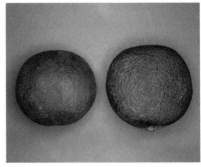

Euphorbiaceae - *Omphalea* (0.66x; 45.16 × 43.71)

Pausandra Radlk. Trees to 10 m tall. Leaves alternate, the margins serrate, the petioles with a prominent pulvinus, the lamina with two glands at base on abaxial surface. Sap reddish. Stipules caducous. In-

Euphorbiaceae - *Pausandra* (2.71x; 8.36 × 6.6)

fructescence axillary. Fruit a trivalved capsule, to 1 cm diameter, brown when mature. Seeds three per fruit. Distribution: Tropical America.

Pera Mutis. Shrubs or trees to 30 m tall. Leaves opposite in some species, always entire; the laminar surface glaucous abaxially in some species due to presence of scales or stellate hairs; some species with inconspicuous laminar glands. Infructescence axillary or borne from the stems. Fruit a trivalved capsule, to 2 cm diameter, brown or reddish when mature. Seeds various per fruit, covered by red aril. Distribution: Tropical and subtropical America.

Euphorbiaceae - *Pera* (6.93×; 4.55 × 3.5)

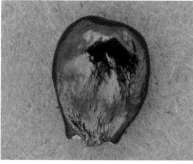

Euphorbiaceae - *Pera* (5.14×; 8.52 × 6.56)

Richeria Vahl. Trees to 25 m tall. Leaves alternate, grouped at the branch apices, the young leaves and branch apices with dense pubescence. Infructes-

Euphorbiaceae - *Richeria* (2.16×; 10.44 × 5.88)

cence axillary. Fruit a trivalved capsule, to 2 cm diameter, yellow when mature. Seeds one per fruit, covered by red or orange aril. Distribution: Tropical and subtropical America.

Ricinus L. Shrubs to 3 m tall. Leaves alternate, lobed, peltate, the margin serrate. Fruit a trivalved capsule, to 2 cm diameter, brown when mature, the exocarp spinose. Seeds 1–3 per fruit. Cultivated for oil extracted from the seeds. Seeds historically used to produce castor oil ingested by humans as medicine, but the toxic chemical ricin is also extracted from the seeds. Ornamental. Distribution: Introduced from Africa and occurring throughout tropical and subtropical America, to 3000 m elevation.

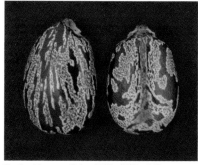

Euphorbiaceae - *Ricinus* (2.33×; 11.32 × 7.68)

Sapium P. Browne. Trees to 30 m tall. Leaves alternate, glabrous, the margins entire or lightly serrate, the secondary venation rather inconspicuous, the petiole or lamina base often with a conspicuous or inconspicuous pair of glands. Latex white, abundant. Infructescence terminal. Fruit a trivalved capsule, to 1.4 cm diameter, green when mature. Seeds 1–3 per fruit with red or orange aril. Genus common in secondary forests and natural clearings. Distribution: Panama and Guianas to Peru and Bolivia, to 1500 m elevation, also in the Old World tropics; introduced and invasive in some warm temperate zones of the southern United States.

Euphorbiaceae - *Sapium* (6.3×; 5.18 × 4.37)

Senefeldera Mart. Trees to 25 m tall. Leaves alternate, weakly grouped at the branch apices; the adaxial laminar surface glossy; the petiole apex with a conspicuous pulvinus near junction with lamina. Latex white, inconspicuous. Infructescence terminal. Fruit a trivalved capsule, to 2.5 cm diameter, brown when mature, the stigma persistent. Seeds three per fruit. Distribution: Guianas to Peru and Bolivia.

Euphorbiaceae - *Senefeldera* (1.99x; 12.61 × 12.71)

FABACEAE

Family diverse, abundant, widespread, and morphologically quite variable. Although treated by some authors with three separate subfamilies, here we recognize the Fabaceae as one family. The family is composed of trees, shrubs, herbs, lianas, and vines. Leaves generally compound, pinnate, bipinnate, and bi- or trifoliate, rarely simple, alternate and rarely opposite; the petiole and petiolules conspicuously cylindrical, swollen in the form of a pulvinus or pulvinulus. Stipules present and sometimes the leaflets with stipels. Fruits usually a legume with one locule and one carpel, but sometimes drupes, berries, loments, or samara. Some genera with red sap, others with glands or spines.

Abarema Pittier. Trees to 25 m tall. Leaves bipinnately compound, the pinnae paripinnate, the leaflets opposite, sometimes rhomboid; glands present along the rachis of each pinnae, just below the junction of each leaflet. Infructescence terminal. Fruit a legume, spiral, to 4 cm diameter, 1–1.5 cm wide, brown outside and red inside when mature. Seeds 10–12 per fruit, often not all seeds are viable. Plants sometimes with short pubescence throughout the leaf, except on the adaxial surface. Genus segregated from *Pithecellobium* . Distribution: Guianas to Peru and Bolivia.

Acacia Mill. Trees to 30 m tall, or lianas, some species with spines on the trunk, and some lianas with curved spines on the stems or leaves. Leaves bipinnately compound, the pinnae paripinnate, the leaflets alternate or opposite, usually quite small; some

species with glands between the last pair of leaflets of each pinnae or between the last 2–4 pairs of leaflets. Stipules axillary. Infructescence terminal or axillary. Fruit a legume, dry, to 25 cm long, brown when mature, some species with whitish very short and dense pubescence. Seeds 6–10 per fruit, remaining connected to the valve by the funiculus after the fruit opens. Plants glabrous or with short pubescence, often growing in disturbed areas, such as secondary forest. Some species cultivated as ornamentals. Distribution: Subtropical and tropical regions of the Americas, Africa, Asia, and Australia, to 3500 m elevation.

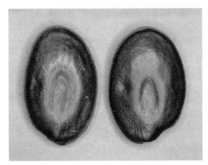

Fabaceae - *Acacia* (5.24x; 7.58 × 4.99)

Fabaceae - *Acacia* (2.12x; 15.3 × 11.85)

Aeschynomene L. Herbs or shrubs to 1 m tall, sometimes with small spines on stems and leaves. Leaves pinnately compound, the leaflets opposite and very small. Infructescence terminal. Fruit a lo-

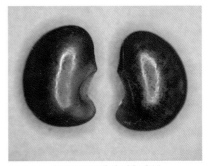

Fabaceae - *Aeschynomene* (9.68x; 3.5 × 2.64)

ment with 3–6 segments, with one flat side and one undulated side, to 4 cm long, black when mature, subtended by persistent calyx and stamens. Seeds one per fruit segment. Many species of disturbed forests. Distribution: Mexico to Peru and Bolivia.

Amburana Schwacke & Taub. Trees to 40 m tall. Leaves compound, paripinnate, the leaflets alternate. Infructescence axillary or terminal. Fruit a bivalved capsule, to 7 cm long, swollen at each seed. Seeds winged, usually one per fruit. Trunk with papyraceous bark. Plants somewhat aromatic. Some species harvested for their wood. Distribution: Brazil, Peru, Bolivia, and Argentina, to 1200 m elevation.

Fabaceae - *Amburana* (1.1×; 55.29 × 14.42)

Apuleia Mart. Trees to 40 m tall, with large flat buttresses. Leaves compound, imparipinnate, the leaflets alternate, some species golden, nearly glaucous abaxially, glabrous adaxially. Stipules sometimes present. Infructescence axillary. Fruit a legume, indehiscent, brown to yellowish-brown with velvet pubescence when mature, to 5 cm long with one conspicuous margin. Seeds 1–2 per fruit. Distribution: Brazil, Peru, Bolivia, and Argentina.

Fabaceae - *Apuleia* (4.28×; 8.24 × 5.87)

Bauhinia L. Trees, shrubs, or lianas. Leaves compound, unifoliate, entire or partially to completely bifid from apex toward base, sometimes conspicuously 3–6-nerved. Stipules caducous. Infructescence axil-

lary or terminal. Calyx persistent, subtending the fruit. Fruit a legume, dry, to 9 cm long, black when mature. Seeds 2–4 per fruit. Some species with short pubescence throughout the plant or with reddish-golden abaxial leaf surface. Other species with tendrils or rarely with spines. Some species of liana with flattened, undulate stem. Distribution: Tropical America, also in the Old World.

Fabaceae - *Bauhinia* (3.51×; 10.15 × 8.22)

Caesalpinia L. Shrubs to 4 m tall, rarely trees or lianas, sometimes with spines. Leaves bipinnate, the pinnae paripinnate, the leaflets opposite. Stipules present at the union of the lower pinnae. Infructescence axillary or terminal. Fruit a legume to 12 cm long, brown when mature, the exocarp sometimes spinose. Seeds various per fruit. Plants used for ornamental purposes. Distribution: Possibly native to Asia, widely cultivated throughout the tropical and subtropical regions of the world, to 2500 m elevation.

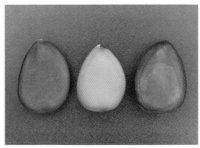

Fabaceae - *Caesalpinia* (2.35×; 10.5 × 7.39)

Calliandra Benth. Trees to 10 m tall. Leaves bipinnately compound, each pinnae with two pairs of leaflets, the principal vein of the leaflet to one side, the secondary vein parallel to principal vein, the petiolule of each leaflet pubescent or glabrous, sometimes with a stipels at its base. Stipules in one obvious pair at leaf base. Infructescence axillary. Fruit a legume to 8 cm long, with prominent margin. Seeds various per fruit. Common along river banks, rare in the forest understory. Plants widely cultivated as ornamentals. Distribution: Mexico to Peru and Bolivia, also in the Old World.

Fabaceae - *Calliandra* (4.89×; 8.74 × 5.82)

Calopogonium Desv. Lianas. Leaves trifoliate. The leaflets weakly asymmetric. Stipules present. Infructescence terminal. Fruit a legume, to 8 cm long, dark brown when mature. Seeds various per fruit. Plants with short, yellow, dense pubescence, light in color on adaxial laminar surface. More common in disturbed areas. Distribution: Mexico to Peru and Bolivia.

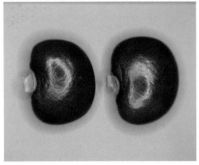

Fabaceae - *Calopogonium* (5.93×; 4.93 × 4.29)

Canavalia DC. Lianas. Leaves trifoliate. Leaflets weakly asymmetric, with stipels at their base. Fruit a legume to 20 cm long, brown when mature with a prominent marginal line on both sides of the fruit. Seeds to 14 per fruit. Plants with smooth pubescence on the stem, petiole, petiolule, and fruit. Distribution: Panama to Argentina, also in the Old World.

Fabaceae - *Canavalia* (1.71×; 22.16 × 12.47)

Cedrelinga Ducke. Trees to 45 m tall. Leaves bipinnate, with 1–2 pairs of opposite, paripinnate pinnae, each terminating in a structure that resembles an apical bud; the leaflets of the pinnae opposite; the petioles glandular. Infructescence axillary or terminal. Fruit a loment to 50 cm long, yellow or brown when mature, the segments 2–6, spiralform. Seeds one per segment of the fruit. The wood is highly valued in the western Amazon. Distribution: Colombia, Brazil, Ecuador, Peru, and Bolivia.

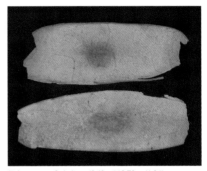

Fabaceae - *Cedrelinga* (0.49×; 119.78 × 41.96)

Centrolobium Benth. Trees to 25 m tall. Leaves compound, paripinnate, the leaflets light pubescent with yellow punctations. Fruit a spinose samara to 25 cm long, brown when mature. Seeds one per fruit. Distribution: Panama to Ecuador and southern Brazil.

Fabaceae - *Centrolobium* (0.33×; 250 × 90)

Fabaceae - *Centrolobium* (0.46×; 250 × 90)

Centrosema (DC.) Benth. Lianas. Leaves trifoliate, with stipules at the base, the leaflets asymmetric with one pair of stipels on the petiolules and with another pair at the base of the central leaflet. The calyx persistent. Fruit a legume, to 15 cm long, brown when mature. Plants pubescent throughout, sometimes golden. Many species used as forage for livestock. Flowers and roots medicinal. Distribution: Tropical and subtropical America, to 3000 m elevation.

Fabaceae - *Centrosema* (4.23x; 8.43 × 5.8)

Clitoria L. Sprawling or climbing herbs, shrubs, or rarely trees. Leaves trifoliate, rarely pinnately compound, with stipules at the base; stipels present at the union of the leaflets, with one pair at the base of the central leaflet. Infructescence axillary. Fruit a legume to 15 cm long, brown when mature, with one prominent longitudinal line, spiralform when dry, the bracts and stigma persistent. Seeds 10 per fruit. Plants with short, dense pubescence on the stems, petioles, and abaxial laminar surface. Plants very common in disturbed areas and along the beaches and banks of rivers. Distribution: Mexico and Guianas to Bolivia and Peru, also in the Old World.

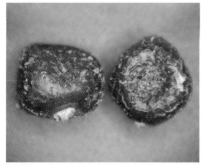

Fabaceae - *Clitoria* (6.72x; 4.54 × 4.27)

Copaifera L. Trees to 45 m tall. Trunk cylindrical or with small buttresses at base. Leaves compound, imparipinnate, the leaflets alternate or subopposite, weakly asymmetric, small, coriaceus and shiny. Infructescence terminal. Fruit a legume to 4 cm long, reddish to black when mature. Seeds 1–2 per fruit,

half covered by a yellow aril. Valued for wood and more for oil that is extracted from trunks. Distribution: Brazil, Peru, and Bolivia.

Fabaceae - *Copaifera* (2.15x; 19.46 × 13.35)

Crotalaria L. Herbs or shrubs to 2 m tall. Leaves trifoliate, rarely simple, the abaxial laminar surface glaucous in some species. Infructescence terminal or axillary, with bracteoles at base of pedicel. Fruit a legume, somewhat inflated, to 4.5 cm long, black when mature, the calyx and stigma persistent. Seeds various per fruit. Plants with whitish pubescence throughout. Common in disturbed forests and along rivers. Distribution: Tropical America, also in the Old World.

Fabaceae - *Crotalaria* (7.51x; 2.45 × 2.39)

Crudia Schreb. Trees to 30 m tall. Leaves compound, imparipinnate with 4-7 alternate leaflets, the stipules caducous. Infructescence terminal, solitary. Fruit a legume, to 13 cm long, brown, tomentose, and

Fabaceae - *Crudia* (1.08x; 45.1 × 45.49)

wrinkled when mature. Seeds 1–2 per fruit. Generally in flooded or seasonally flooded forests or at the edge of oxbow lakes. Distribution: Guianas to Peru and Bolivia.

Dalbergia L. f. Lianas or shrubs to 5 m tall, some species with spines. Leaves compound, imparipinnate, with 1–15 alternate, rarely subopposite, cordate leaflets; stipules caducous. Infructescence terminal or axillary. Fruit a samara, to 3.5 cm diameter, green or brown when mature. Seeds one per fruit or one per segment. Plants pubescent or glabrous. Distribution: Southern United States to Peru and Bolivia, also in the Old World.

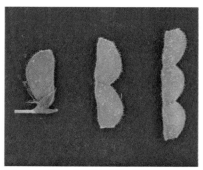

Fabaceae - *Desmodium* (3.3x; 5.64 × 2.98)

Fabaceae - *Dalbergia* (1.78x; 34.17 × 9.29)

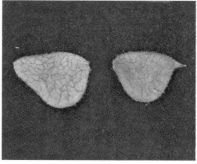

Fabaceae - *Desmodium* (3.72x; 6.99 × 4.91)

Desmodium Desv. Erect or sprawling herbs, or lianas. Leaves compound, trifoliate, with axillary stipules; the leaflets often asymmetric, with a pair of stipels at the union of the lateral leaflets and a pair of stipels at the base of the terminal leaflet. Infructescence terminal or axillary. Fruit a loment, to 10 cm long, with one to multiple segments, the surface asperous permitting the fruit to adhere to other surfaces (e.g., mammal hair and clothing); brown when mature and subtended by persistent calyx. Seeds one per segment. Plants with short pubescence throughout, except on adaxial laminar surface; more dense and conspicuous on the abaxial surface. All parts of the plant medicinal. Plants of secondary forest and river margins. Distribution: Mexico to Peru and Bolivia, to 3500 m elevation.

Dialium L. Trees to 30 m tall. Leaves compound, imparipinnate, with 3–5 alternate leaflets; the stipules caducous. Sap reddish, very inconspicuous from the trunk. Infructescence terminal or axillary. Fruit a drupe, to 2 cm long, brown, short pubescent when mature, and subtended by persistent calyx in the form of a ring; the exocarp thin. Seeds one per fruit, usually free, making noise when shaken. Distribution: Guatemala to Peru and Bolivia, but more speciose in the Old World.

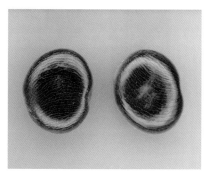

Fabaceae - *Dialium* (2.9x; 9.82 × 8.39)

Dioclea H.B.K. Lianas. Leaves compound trifoliate, with stipules at base of petiole; sometimes with a pair of stipels at the union of the lateral leaflets and on the terminal petiolule. Infructescence axillary. Fruit a

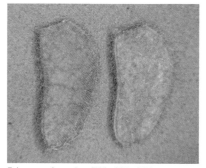

Fabaceae - *Desmodium* (7.25x; 5.96 × 2.57)

legume to 25 cm long, green or brown when mature. Seeds to 10 per fruit. Some species with short, dense, sometimes golden pubescence throughout, except on the adaxial leaf surface. Usually in secondary forest, along the margins of rivers, or along small streams. Distribution: Mexico to Peru and Bolivia, also in Africa and Asia.

Fabaceae - *Dioclea* (1.11×; 33.05 × 26.76)

Fabaceae - *Dioclea* (3.68×; 11.26 × 6.44)

Diplotropis Benth. Trees to 30 m tall. Leaves compound, imparipinnate, with 3–15 alternate leaflets. Infructescence terminal or axillary. Fruit variable, round, and woody, or a membranous legume, to 15 cm long, brown when mature, subtended by persistent calyx. Seeds 1–3 per fruit. Distribution: Guianas to Peru and Bolivia.

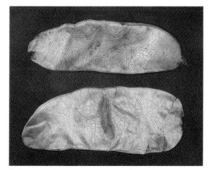

Fabaceae - *Diplotropis* (0.56×; 99.43 × 36.38)

Dipteryx Schreb. Trees to 55 m tall. Trunk with large buttress roots at base. Leaves compound, imparipinnate, with 5–10 leaflets that are alternate at the apex or subopposite to opposite at the base; the leaf rachis flattened and weakly winged, the principal vein of the leaflets situated to one side. Infructescence terminal. Fruit a drupe to 7 cm long, brown when mature, the mesocarp green, the endocarp woody, fibrous. Seeds one per fruit. Distribution: Costa Rica to Peru and Bolivia.

Fabaceae - *Dipteryx* (1.13×; 51.21 × 34.11)

Dussia Krug & Urb. ex Taub. Trees to 35 m tall. Leaves compound, imparipinnate with 7–15 alternate, subopposite, to opposite leaflets; the tertiary veins parallel to one another and perpendicular to the secondary veins. Some species with reddish sap. Infructescence terminal or axillary. Fruit a legume, to 10 cm long, orange or brownish-orange and pubescent when mature; subtended by persistent calyx. Seeds 1–2 per fruit, the surface black and the interior white, covered by red aril. Plants with short, white pubescence on the abaxial laminar surface; very dense, rufescent to orangish tomentose pubescence on the petioles and terminal regions of the branches. Distribution: Southern Mexico and the Antilles to Peru and Bolivia.

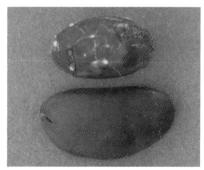

Fabaceae - *Dussia* (1.71×; 27.1 × 13.95)

Entadopsis Adans. Lianas and trees. Stems lined with obscure longitudinal striations. Leaves bipinnately compound, the pinnae paripinnate in 4–6 pairs; the leaflets opposite, nearly sessile, with asymmetric bases; the principal vein of the leaflets weakly oriented to one side. Some species with tendrils. Infructescence axillary or terminal. Fruit a loment or legume to 100 cm long, brown or black when mature, with up to 15 segments that separate leaving the intact margin of the whole fruit. Plants with short, yellow pubescence on the leaves, pedicel, and terminal region of the branches. Some species with edible leaves and seeds. Distribution: Tropical America and Africa.

Fabaceae - *Entadopsis* (0.87×; 71.34 × 18.66)

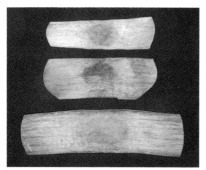

Fabaceae - *Entadopsis* (0.64×; 87.6 × 22.15)

Enterolobium C. Mart. Trees to 40 m tall. Leaves bipinnately compound, the leaflets of the pinnae sessile, opposite, and usually small; glands present on the petiole, principal rachis, and rachis of each pinnae. Stipules caducous. Infructescence axillary or terminal. Fruit a kidney-shaped or spiralform legume, to 13 cm diameter, usually black when mature; the mesocarp white, mealy. Seeds various per fruit, usually free in each segment of the fruit. Some species with very short pubescence throughout the plant. Trees valued for lumber. The bark used to produce soap, medicine, and tannins. The seeds are used in arts and crafts. Distribution: Mexico to Peru and Bolivia, to 1800 m elevation.

Fabaceae - *Enterolobium* (2.03×; 22.55 × 10.73)

Fabaceae - *Enterolobium* (2.5×; 10.04 × 7.54)

Erythrina L. Trees to 30 m tall, rarely shrubs. Trunks with spines when young. Leaves trifoliate, glandular, with stellate pubescense; stipules present at the base of petioles, and stipels on the leaflets. Infructescence axillary or terminal. Fruit a legume to 20 cm long, compressed on each seed, brown when mature, subtended by persistent calyx. Seeds various per fruit, some species with one seed per globose fruit. Caducifolius. Plants often used as forage, ornamentals, or living fences. Widely cultivated throughout the tropics. Distribution: Central America to Bolivia and Peru, also in the Old World, to 3300 m elevation.

Fabaceae - *Erythrina* (2.41×; 9.78 × 6.45)

Fabaceae - *Erythrina* (2.88×; 13.09 × 5.37)

Hymenaea L. Trees to 40 m tall. Leaves compound, bifoliate; the leaflets coriaceous with asymmetric bases; the principal vein of the leaflets curved and positioned slightly to one side. Stipules caducous. Infructescence terminal. Fruit a legume to 16 cm long, brown when mature; the exocarp woody; the mesocarp white, mealy, sweet, edible. Seeds 1–4 per fruit. Some species pubescent. Plants valued for their wood and in some tropical cities the fruits are commercially available in markets. Distribution: Mexico to Paraguay, with one species in Africa, to 1300 m elevation.

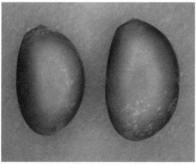

Fabaceae - *Hymenaea* (1.6×; 25.56 × 15.89)

Fabaceae - *Hymenaea* (1.47×; 25.04 × 26.83)

Inga Mill. Shrubs or trees to 30 m tall. Leaves compound, paripinnate, with 1–8 pairs of opposite leaflets; the rachis sometimes winged; glands present

between each pair of leaflets, making it one of the easiest genera to recognize in the Neotropics. Stipules caducous, sometimes quite conspicuous. Infructescence axillary or terminal. Fruit a legume to 50 cm long, straight, curved, or spiraled, flat or round in cross section, sometimes only partially dehiscent, green or yellow when mature. Seeds various per fruit, with a white, mealy, often sweet aril. Plants pubescent or glabrous. Some species valued for wood, but usually used as firewood and to build fences. Plants also grown as shade for cacao and coffee plantations. Widely cultivated for edible fruits, which are often found in tropical fruit markets (i.e., pacay and guava). Distribution: Mexico to Peru and Bolivia, to 2400 m elevation.

Fabaceae - *Inga* (2.31×; 13.12 × 6.93)

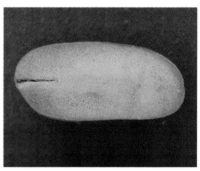

Fabaceae - *Inga* (3.75×; 15.56 × 7.03)

Fabaceae - *Inga* (2.41×; 9.39 × 6.14)

Lecointea Ducke. Trees to 30 m tall. Leaves simple, the margin weakly to strongly serrate, the venation rather inconspicuous. Stipules caducous. Infructescence axillary. Fruit a drupe to 6 cm long, white and aromatic when mature. Seeds 1–2 per fruit. Plants glabrous. Distribution: Honduras to Peru and Bolivia.

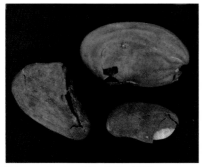

Fabaceae - *Lecointea* (1.13×; 33.18 × 18.22)

Machaerium Pers. Usually lianas, rarely shrubs and trees; some species with spines developed from modified stipules. Leaves compound, imparipinnate with five to more than 50 alternate leaflets. Plants usually with red sap. Infructescence terminal or axillary. Fruit a samara to 10 cm long, the wing curved or straight, venation reticulate, sometimes with protuberances on the seed end, brown when mature, subtended by persistent calyx. Seeds one per fruit. Pubescence brown or white, very dense throughout the plant except on the adaxial laminar surface. Distribution: Tropical America to 1500 m elevation.

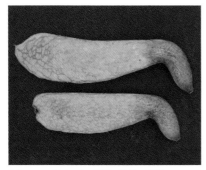

Fabaceae - *Machaerium* (1.05×; 56.67 × 13.74)

Macrosamanea Britton et Rose. Shrubs or trees to 30 m tall. Leaves bipinnately compound, the pinnae paripinnate, leaflets opposite, nearly sessile, asymmetric, the leaf rachis canaliculate with one pair of glands near the base and a second between the last pair of pinnae. Fruit a legume to 10 cm long, black when mature. Seeds 10–12 per fruit. Plants with short pubescence sometimes white or yellowish throughout. Genus segregated from *Pithecellobium* . Distribution: Guianas and Surinam to Peru and Bolivia.

Fabaceae - *Macrosamanea* (6.26×; 7.25 × 3.06)

Mimosa L. Erect or prostrate-spreading herbs, shrubs, or lianas, generally with spines on stems and trunk. Leaves bipinnately compound, sometimes glandular, with one or more than 20 pairs of pinnae; some species with sensitive leaflets that close when touched. Stipules present at base of leaves. Infructescence axillary or terminal. Fruit a loment, to 18 cm long, separating into segments leaving an intact marginal connecting vein, brown when mature, sometimes with spines. Seeds one per segment. Plants glabrous or pubescent. Distribution: Central and South America, also in Africa and Asia, to 3000 m elevation.

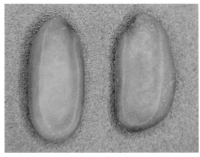

Fabaceae - *Mimosa* (8.62×; 4.88 × 2.47)

Mucuna Adans. Lianas. Leaves compound, trifoliate. Stipules and stipels present. Infructescence axillary. Fruit a legume to 16 cm long, brown or black when mature, some species with urticating hairs.

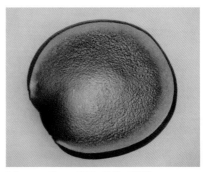

Fabaceae - *Mucuna* (1.25×; 41.02 × 37.03)

Seeds 1–3 per fruit. Plants glabrous or pubescent, sometimes golden. Widely cultivated as forage and for improving soils. Distribution: Tropical America and Africa.

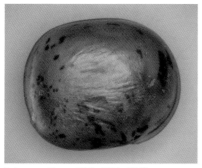

Fabaceae - *Mucuna* (1.22×; 44.26 × 35.65)

Fabaceae - *Mucuna* (1.59×; 30 × 27.73)

Fabaceae - *Mucuna* (1.93×; 30.44 × 21.04)

Myroxylon L. f. Trees to 35 m tall. Leaves compound, imparipinnate, the leaflets alternate with inconspicuous translucent punctations. Infructescence axillary or terminal. Fruit a samara to 10 cm long, brownish-yellow when mature. Seeds one per fruit. Some parts of the plant with a balsam odor, sometimes very aromatic. Trees valued for wood and the resin used to make balsam that is available in commercial markets. Distribution: Mexico to Bolivia, to 2200 m elevation.

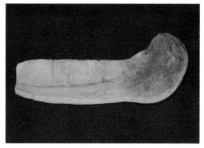

Fabaceae - *Miroxylon* (0.66×; 85.48 × 22.9)

Ormosia Jacks. Trees to 35 m tall. Leaves compound, imparipinnate with 5–13 opposite or subopposite leaflets. Some species with red sap in the stems or with stipules. Infructescence axillary or terminal. Fruit a legume, to 8 cm long, subwoody, brown or reddish-brown when mature; subtended by persistent calyx. Seeds 1–5 per fruit. Some species with short pubescence, brown or yellow throughout the plant, except on the adaxial laminar surface. Trees valued for lumber. Seeds used in arts and crafts. Distribution: Tropical America, Asia, and Australia, to 2000 m elevation.

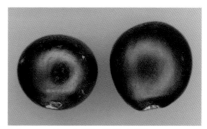

Fabaceae - *Ormosia* (2.2×; 13.65 × 13.03)

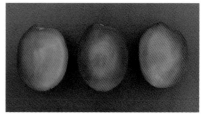

Fabaceae - *Ormosia* (2.12×; 11 × 8.3)

Fabaceae - *Ormosia* (1.88×; 15.79 × 13.93)

Fabaceae - *Ormosia* (3.81×; 10.81 × 9.22)

Parkia R. Br. Trees to 40 m tall. Leaves bipinnately compound, sometimes opposite but often clustered at the branch apex, with 3-16 pairs of opposite or alternate paripinnate pinnae; the leaflets numerous, opposite and sessile, generally very small; glands present on petioles and in one pair at the union of the 2–3 basal pairs of leaflets. Infructescence terminal. Fruit a legume to 40 cm long, black when mature, weakly curved, 1–10 per infructescence, sometimes long-pedicillate. Seeds 15–25 per fruit, each surrounded by a sticky aril. Some species with a wide, thin, flat canopy. Plants glabrous or pubescent. Trees valued for their wood. Distribution: Tropical America, Africa, and Asia.

Fabaceae - *Parkia* (1.64×; 21.68 × 9.29)

Fabaceae - *Parkia* (3.92×; 11.27 × 5.32)

Phyllocarpus Riedel ex. Endl. Trees to 30 m tall. Leaves compound, paripinnate, with five opposite pairs of leaflets. Infructescence axillary. Fruit a membranaceous pod to 12 cm long, with a submarginal rib along the wing-like margin. Seeds 1–2 per fruit. Plants caducifolious. Distribution: Tropical America.

Fabaceae - *Phyllocarpus* (0.57×; 107.47 × 39.45)

Pseudopiptadenia Rauschert. Trees to 35 m tall. Leaves bipinnately compound with 10–13 pairs of opposite to subopposite paripinnate pinnae, and opposite, sessile leaflets. Fruit a legume to 30 cm long, brown when mature, the margin prominent. Seeds many per fruit, winged. Distribution: Tropical America.

Fabaceae - *Pseudopiptadenia* (1.42×; 43.21 × 11.57)

Pterocarpus Jacq. Shrubs or trees to 40 m tall. Leaves compound, imparipinnate, the leaflets 5–8, alternate; the margin weakly serrate; the venation very conspicuous; rarely trifoliate. Stipules present. Sap reddish in the wood. Infructescence axillary or terminal. Calyx persistent. Fruit a samara to 8 cm diameter, brown or yellow when mature, flat, winged; the wing undulate, completely surrounding the seed. Seeds one or rarely two per fruit. Trees valued for wood. Distribution: Tropical America, also in Africa and Malesia.

Rhynchosia Lour. Lianas. Leaves compound, trifoliate, the leaflets with glands and stipels. Infructescence axillary. Fruit a capsule, to 10 cm long. Seeds one to various per fruit, used in arts and crafts. Distribution: Southern United States to all of tropical America, but more speciose in the Old World.

Schizolobium Vogel. Trees to 30 m tall. Leaves bipinnately compound, the pinnae paripinnate, opposite, with numerous tiny, opposite leaflets. Caducifolius. Fruit a legume to 10 cm long, brown when mature, reticulately veined, subwoody, the two valves dehiscing completely along their entire length. Seeds one per fruit, winged. Trees valued for wood and cultivated for their rapid growth and ornamental value. Distribution: Central America to Peru and Bolivia.

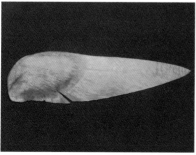

Sclerolobium Vogel. Trees to 25 m tall. Leaves compound, paripinnate, with 8–11 pairs of opposite leaflets and large stipules sometimes resembling stunted leaves at the petiole base; the petioles swollen in some species, often harboring ants. Infructescence terminal. Fruit a legume to 10 cm long, the exocarp defoliating, brown when mature. Seeds 1–2 per fruit. Pubescence short tomentose on the abaxial leaflet surface, petiole, and branch apex. Similar to *Tachigali* . Distribution: Tropical America, to 1500 m elevation.

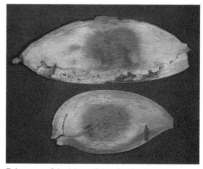

Senna Miller. Lianas, shrubs, or trees to 15 m tall. Leaves compound, paripinnate, with 2–6 pairs of opposite leaflets; one or two glands prominent on the rachis, between the leaflets; some species

with a pair of curved spines at the base of each leaf. Stipules axillary, resembling stunted leaves. Infructescence terminal or axillary. Fruit a legume to 25 cm long, brown when mature. Genus segregated from *Cassia* . Distribution: Cosmopolitan on five continents, to 4000 m elevation.

Swartzia Schreber. Trees to 40 m tall. Leaves compound, imparipinnate, with 3–11 opposite leaflets; the rachis winged in some species; rarely unifoliate. Stipules caducous, stipels present. Some species with reddish sap in the wood. Infructescence axillary or borne from the trunk. Fruit a legume to 14 cm long, smooth or tuberculate, fleshy to subwoody, yellow when mature. Seeds 1–4 per fruit, with a white, cream, or yellow aril and an elongate funicule. Pubescence short, dense in some species. Distribution: Tropical America, also in Africa.

Fabaceae - *Swartzia* (1.46×; 19.06 × 14.42)

Tachigali Aubl. Trees to 30 m tall. Leaves compound, paripinnate, with 8–12 pairs of opposite leaflets; usually terminating in an aborted bud. Fruit a legume to 20 cm long, brown when mature. Seed 1–2 per fruit. Many species monocarpic. The abaxial surfaces of the leaflets golden or glaucous in some species, pubescent or glabrous in others. Some species with myrmecophilous petioles, harboring ants. In some texts written as *Tachigalia* . Distribution: Costa Rica to Peru and Bolivia.

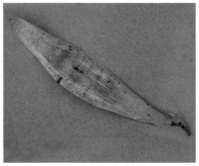

Fabaceae - *Tachigali* (0.76×; 86.39 × 17.92)

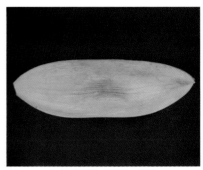

Fabaceae - *Tachigali* (0.47×; 131.24 × 40.45)

Vatairea Aubl. Trees to 20 m tall. Leaves compound, imparipinnate, with 15–19 alternate to subopposite leaflets; the margins lightly serrate. Stipules caducous. Infructescence terminal. Fruit a samara to 11 cm long, or a drupe to 9 cm diameter, brown when mature, subtended by persistent calyx. Seeds one per fruit. Pubescence short and dense on the stem and petiole. Distribution: Mexico to Peru and Bolivia.

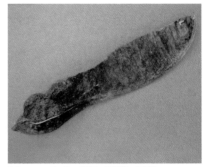

Fabaceae - *Vatairea* (0.72×; 100.88 × 17.79)

Vigna Savi. Lianas. Leaves trifoliate. Stipules obvious, stipels short. Fruit a capsule, to 8 cm long, brown when mature. Seeds various per fruit. Pubescence short, light-colored throughout the plant. The genus easily confused with *Phaseolus* . Distribution: Pantropical.

Fabaceae - *Vigna* (7.88×; 3.89 × 2.86)

Zygia P. Browne. Trees to 15 m tall. Leaves bipinnately compound, the pinnae paripinnate, with 4–6 alternate to subopposite leaflets on each pinnae, the terminal pair opposite; the pinnae sometimes with one gland between the terminal or last two pairs of leaflets. Infructescence borne from the stems or trunk. Fruit a legume to 15 cm long, purple or reddish when mature. Seeds various per fruit. Genus segregated from *Pithecellobium* . Distribution: Tropical America.

Fabaceae - *Zygia* (1.25×; 20.34 × 16.95)

Fabaceae - *Zygia* (1.33×; 14.6 × 13.57)

FLACOURTIACEAE

Shrubs and trees. Leaves simple, alternate, one genus with opposite leaves, the margins usually serrate. Stipules persistent or caducous. Some genera with translucent punctations on the leaves, glands or spines, but otherwise lacking vegetative characteristics, making generic identification difficult in sterile condition. Fruit almost always a capsule, sometimes a berry. Seeds few to numerous, with or without an aril.

Banara Aubl. Shrubs or trees to 10 m tall. Leaves coriaceous with serrate margins, a pair of glands on the adaxial surface near the junction of the petiole, and sometimes with glands on the teeth of dentate leaf margins. Stipules caducous. Infructescence terminal. Calyx and style persistent. Fruit a berry to 1.5 cm diameter, yellow when mature. Seeds many per fruit. Plants of disturbed zones or areas of abundant light. Distribution: Mexico to Chile to 2100 m elevation.

Flacourtiaceae - *Banara* (12.17×; 3.35 × 2.01)

Carpotroche Endl. Trees to 10 m tall. Leaves usually large, the margins serrate, the petiole with a prominent pulvinus at the base of the leaf. Infructescence borne from the trunk. Fruit a 3- to 5-valved capsule, to 7 cm, white to greenish when mature; the exocarp with lightly spinose protuberances. Seeds many per fruit, covered partially with an orange aril. Distribution: Guianas to Peru.

Flacourtiaceae - *Carpotroche* (2.81×; 9.92 × 4.89)

Casearia Jacq. Shrubs or trees to 30 m tall. Leaves with serrate margins, and sometimes with linear or round, translucent punctations. Stipules persistent or caducous. Infructescence axillary or borne from the stems or trunk. Fruit generally a capsule, usually trivalved, sometimes an indehiscent berry, to 10 cm diameter, red, yellow, orange, or purple to brown when mature; the calyx and stigma persistent in some species. Seeds one to many per fruit, generally embedded in a juicy, transparent or yellowish mesocarp, sometimes covered partially by a white aril. Plants pubescent or glabrous, rarely with spines. Distribution: Mexico and Puerto Rico to Uruguay.

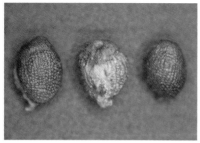

Flacourtiaceae - *Casearia* (11.85×; 1.92 × 1.39)

Flacourtiaceae - *Casearia* (6.72×; 3.94 × 3.83)

Flacourtiaceae - *Casearia* (4.91×; 8.62 × 4.89)

Hasseltia H.B.K. Shrubs to 15 m tall. Leaves with serrate margin on the upper half and clearly trinerved from the base, with one pair of glands at the base of the lamina on the adaxial surface. Stipules caducous. Infructescence axillary or terminal. Fruit a

Flacourtiaceae - *Hasseltia* (5.13×; 5.57 × 4.08)

berry to 1 cm diameter, dark purple or black when mature, with persistent calyx, stamens, and stigma. Seeds 1–4 per fruit. Pubescence short. Distribution: Mexico to Bolivia and Peru.

Laetia Loefl. ex. L. Trees to 30 m tall. Leaves with finely undulate-crenate margin, the base weakly cordate and asymmetric; some species with conspicuous translucent punctations. Infructescence axillary or terminal. Calyx persistent. Fruit a trivalved capsule, to 2.5 cm long, yellow-red when mature. Seeds many per fruit, with white aril. Distribution: Central America to Peru and Bolivia.

Flacourtiaceae - *Laetia* (6.67×; 4.03 × 3.49)

Lindackeria C. Presl. Trees to 12 m tall. Leaves with pulvini at both extremes of the petiole, but more conspicuous at the base of the lamina; some species with conspicuous translucent punctations. Stipule caducous. Infructescence axillary. Fruit a trivalved capsule, to 2.5 cm diameter, yellow when mature, shortly spinose or with protuberances. Seeds various per fruit, covered with a red aril. Distribution: Central and South America, also in Africa.

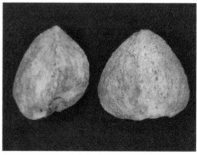

Flacourtiaceae - *Lindackeria* (4.11×; 7.68 × 7.42)

Lunania Hook. Trees to 10 m tall. Leaves conspicuously trinerved, the secondary venation perpendicular to the midvein, the translucent punctations only slightly conspicuous on lamina. Infructescence terminal. Fruit a trivalved capsule, to 1.0 cm diameter, dark purple or black when mature. Seeds various per fruit covered with an orange aril. Distribution: Primarily in the Antilles, also in South America.

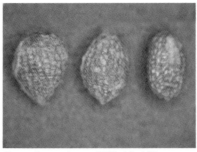

Flacourtiaceae - *Lunania* (13.97x; 1.8 × 1.32)

Mayna Aubl. Shrubs and trees to 15 m tall. Leaves with weakly to strongly serrate margins, or sometimes short spinose in some species; the petioles with pulvini at both extremes; in some species, the lamina with sparse but conspicuous orange translucent punctations. Stipules caducous. Infructescence axillary or borne from the trunk. Fruit a trivalved capsule, to 3.5 cm diameter, yellow when mature; the exocarp weakly spinose; the mesocarp orange or white. Seeds 2–11 per fruit. Plants pubescent or glabrous. Distribution: Central America to Peru and Bolivia, also in Africa, to 1600 m elevation.

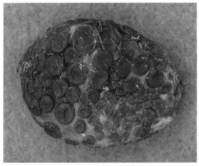

Flacourtiaceae - *Mayna* (4.66x; 12.09 × 9.23)

Flacourtiaceae - *Mayna* (2.33x; 8.5 × 6.59)

Prockia P. Browne ex. L. Shrubs to 3 m tall. Leaves cordate with strongly serrate margins. Stipules caducous or persistent. Infructescence terminal. Calyx and stigma persistent, the first covering two-thirds of the fruit. Fruit a berry, to 1.5 cm diameter, green when mature. Various seeds per fruit. Pubescence

short and white throughout the plant. Glands present on the lamina. Distribution: Central America to Peru and Bolivia, to 2600 m elevation.

Flacourtiaceae - *Prockia* (16.08x; 1.57 × 1.08)

Ryania Vahl. Shrubs to 8 m tall. Leaves with the margin dentate or serrate. Stipules caducous or persistent. Infructescence axillary. Fruit a 5-valved capsule, to 4.5 cm diameter, reddish or dark purple when mature, spongy, subtended by a persistent calyx, the mesocarp white. Seeds many per fruit. Pubescence of short, stellate trichomes, like scales. Distribution: Nicaragua and Guianas to Peru and Bolivia.

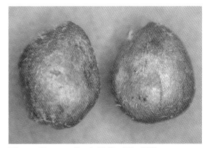

Flacourtiaceae - *Ryania* (7.46x; 4.7 × 3.79)

Xylosma G. Forst. Trees to 30 m tall. Trunk and stems with simple or branched thorns. Leaves with serrate margins. Infructescence axillary. Fruit a berry, to 0.5 cm diameter, red when mature. Seeds few per fruit. Distribution: Tropical and subtropical America, also in Africa, Asia, and Malesia, to 2800 m elevation.

Flacourtiaceae - *Xylosma* (7.09x; 4.72 × 3.16)

GENTIANACEAE

Herbs, saprophytic in some genera, rarely shrubs. Leaves simple and opposite, sometimes verticillate, conspicuously perfoliate, sheathing the stem. Fruit a capsule.

Irlbachia Mart. Herbs to 1 m tall. Stems hollow. Leaves entire, the base perfoliate, completely sheathing the stem and forming a ring; the lamina very thin, nearly transparent when dry. Infructescence terminal or axillary. Calyx and stigma persistent. Fruit a septicidal capsule, to 2 cm long, brown when mature. Seeds many per fruit. Distribution: Tropical America.

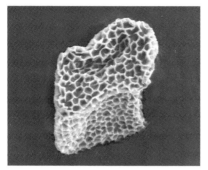

Gentianaceae - *Irlbachia* (751.32x; 0.06 × 0.05)

GESNERIACEAE

Herbs, vines, and hemiepiphytes. Leaves simple, opposite, generally succulent and lightly to strongly pubescent, the margins serrate in many genera. Fruit a capsule or berry. Some genera associated with ants.

Besleria L. Herbs and shrubs. Leaves entire. Infructescence axillary. Fruit a berry to 1 cm diameter, orange, yellow, or red, and in some species opening when mature. Seeds tiny, many per fruit. Plants often pubescent. Common in disturbed areas of forest. Distribution: Tropical America.

Gesneriaceae - *Besleria* (118.52x; 0.4 × 0.38)

Codonanthe (Mart.) Hanst. Epiphytic herbs, rarely shrubs. Leaves carnose, thick succulent. Fruit a berry to 1 cm diameter, red, pink, purple, or black when mature, the mesocarp juicy. Seeds few per fruit. Many species associated with ant gardens. Distribution: Tropical America.

Drymonia Mart. Lianas and climbing herbs, rarely erect. Infructescence axillary. Fruit a berry to 2 cm long, yellow-reddish or purple-black when mature, subtended by persistent red, pink, or green calyx lobes completely sheathing the pedicel and fruit. Distribution: Mexico to Peru and Bolivia.

Gesneriaceae - *Drymonia* (48.68x; 0.89 × 0.44)

GNETACEAE

Lianas. Leaves simple, opposite, and entire, generally coriaceous. Some species exude a sticky cream or transparent resin. The tertiary venation sometimes lightly conspicuous. Family of gymnosperms, not producing fruit, only a naked seed.

Gnetum L. Lianas. Seed solitary, naked, to 5 cm long, red, orange, or purple-black when mature, the inner tissue yellow, fibrous, edible. Distribution: Panama to Peru and Bolivia, but more speciose in the Old World tropics.

Gnetaceae - *Gnetum* (1.02x; 45.62 × 27.5)

HAEMODORACEAE

Herbs. Leaves simple, alternate, succulent, the veins parallel. Fruit a trivalved capsule.

Xiphidium Aubl. Herbs to 0.7 m tall, rhizomatous. Leaves lanceolate, to 70 cm long, the margins spinose, more conspicuous on the young leaves and towards the apex; the base of the leaf sheathing the stem; the venation fine, parallel. Infructescence terminal. Calyx and stigma persistent. Fruit a capsule, to 1 cm long, red when mature. Pubescence short on the stem. Distribution: Tropical America.

Haemadoraceae - *Xiphidium* (59.26x; 0.81 × 0.77)

HELICONIACEAE

Perennial herbs. Leaves organized in a spiral forming a rosette; the petioles with a tubular sheath forming a pseudostem; the central vein prominent; the lamina tearing along the lateral nerves. Infructescence terminal. Bracts of showy colors—reds, yellows, pinks—and attractive to hummingbirds. Family segregated from the Musaceae.

Heliconia L. Herbs to 2 m tall. Leaves to 110 cm long, the petiole to 100 cm long; the secondary and tertiary venation parallel and closely grouped, nearly perpendicular to the principal vein at the base and at 45° angle at the apex. Infructescence erect or pendulant. Fruit sessile, to 2.4 cm long, blue when mature, sub-

Heliconiaceae - *Heliconia* (2.9x; 18.91 × 7.12)

tended by persistent orange, red, yellow, or green bracts; the mesocarp white. Seeds 2–3 per fruit. Widely cultivated as ornamentals and used in cut flower industry. Distribution: Mexico to Peru and Bolivia.

Heliconiaceae - *Heliconia* (3.2x; 13.61 × 8.48)

Heliconiaceae - *Heliconia* (3.68x; 11.87 × 5.83)

HERNANDIACEAE

Trees, shrubs, and lianas. Leaves simple, entire, or parted, alternate, strongly trinerved from the base. Fruit a drupe. Plants aromatic.

Sparattanthelium Mart. Lianas. Leaves conspicuously trinerved from the base, the secondary venation nearly perpendicular to the principal vein. Infructescence axillary or terminal, often falling intact, conspicuous due to white color. Fruit a drupe, to 2

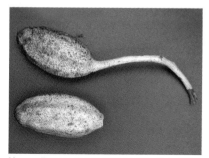

Hernandiaceae - *Sparattanthelium* (1.59x; 19.4 × 9.2)

cm long, white or gray when mature. Pubescence short on the veins of the abaxial laminar surface. Sometimes strongly aromatic when dry. Distribution: Guianas to Peru.

HIPPOCRATEACEAE

Lianas, rarely shrubs or small trees. Leaves simple, opposite, entire. Stipules caducous or absent. Infructescence usually axillary. Fruit a capsule, bi- to trivalved with winged seeds, or a berry.

Anthodon R. & P. Lianas. Leaves with the margin lightly undulate, the venation decurrent. Fruit a trivalved capsule, to 15 cm diameter, the exocarp woody, brown when mature. Seeds winged, many per fruit. Distribution: Panama to Peru.

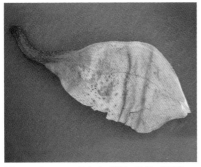

Hippocrateaceae - *Anthodon* (1.26x; 34.62 × 24.37)

Cheiloclinium Miers. Lianas or trees to 8 m tall. Leaves with finely serrate margins. Infructescence axillary or borne from the trunk. Fruit a berry to 7 cm long, orange or yellow when mature, solitary or in pairs; the mesocarp yellow or orange, juicy. Seeds 2–6 per fruit. Distribution: Central America to Paraguay.

Hippocrateaceae - *Cheiloclinium* (1.63x; 23.9 × 11.82)

Hippocrateaceae - *Cheiloclinium* (1.46x; 32.61 × 16.3)

Hippocratea L. Lianas. Leaves with serrate-crenate margins. Fruit a capsule, bivalved, to 7 cm long, brown when mature; the exocarp woody. Seeds winged, 5–7 per fruit. Distribution: Southern United States and Mexico to Peru.

Hippocrateaceae - *Hippocratea* (1.17x; 43.42 × 15.59)

Peritassa Miers. Lianas, rarely shrubs. Leaves in some species alternate, the margins weakly serrate. Infructescence axillary. Fruit a berry, to 8 cm long, striate, orange or yellow when mature, gray-blue when dry; the mesocarp yellow, orange, or greenish. Seeds 2–6 per fruit. Distribution: Panama to Paraguay.

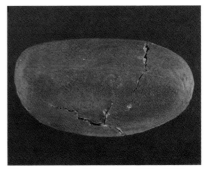

Hippocrateaceae - *Peritassa* (1.58x; 38.99 × 20.74)

Pristimera Miers. Lianas. Leaves with weakly serrate margins. Fruit a bivalved capsule, to 9 cm long, brown when mature; the exocarp woody. Seeds winged, various per fruit. Distribution: Central America and Guianas to Peru.

Hippocrateaceae - *Pristimera* (1.05x; 54.24 × 19.6)

Salacia L. Lianas or trees to 10 m tall. Leaves in some species with weakly undulate margins or the imperceptible tertiary venation. Infructescence axillary or borne from the stems. Fruit a berry, to 18 cm diameter, green, yellow, or orange when mature; the exocarp clearly marked with three longitudinal lines in some species; the mesocarp fleshy, yellow, cream, brown, or orange. Seeds up to 16 per fruit. Distribution: Panama and Guianas to Peru and Bolivia, but more speciose in Africa, to 2500 m elevation.

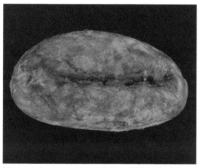

Hippocrateaceae - *Salacia* (1.24x; 49.15 × 26.84)

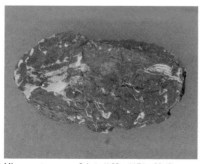

Hippocrateaceae - *Salacia* (1.33x; 44.76 × 23.65)

Tontelea Aubl. Lianas. Leaves with nearly imperceptible tertiary venation. Fruit a berry, to 9.5 cm long, orange or brown when mature; the exocarp lenticellate; the mesocarp orange. Seeds up to five per fruit. Distribution: Panama and Guianas to Peru.

Hippocrateaceae - *Tontelea* (1.49x; 40.14 × 20.21)

HUGONIACEAE

Trees. Leaves simple, alternate, sometimes opposite, always entire. Stipules persistent or caducous. Fruit a drupe with one seed per fruit, or a capsule. Plants glabrous. Recently segregated from the family Linaceae.

Hebepetalum Benth. Trees to 28 m tall. Leaves alternate, the venation with submarginal collecting vein, the border crenate. Infructescence axillary or terminal. Calyx persistent. Fruit a drupe, to 1.5 cm long, black when mature. Seeds one per fruit. Distribution: Tropical America.

Hugoniaceae - *Hebepetalum* (2.98x; 9.4 × 6.49)

Roucheria Planchon. Trees to 20 m tall. Leaves with secondary venation very fine and nearly perpendicular to the midvein, completely anastomosing at 0.2 cm from the border, with a conspicuous lighter-colored band parallel to the midvein (similar to Erythroxylaceae). Stipules caducous. Infructescence terminal. Fruit a drupe to 1.5 cm long, black when mature, subtended by persistent calyx; the mesocarp green-yellow. Distribution: Tropical America, to 1600 m elevation.

Hugoniaceae - *Roucheria* (4.19×; 10.43 × 5.38)

HUMIRIACEAE

Trees. Wood reddish, sometimes resinous. Leaves simple, alternate, generally coriaceous; the margins more or less crenate or serrate in some genera, generally coriaceous. Stipules caducous or absent. Infructescence axillary or terminal. Fruit a drupe.

Sacoglottis Mart. Trees to 30 m tall. Leaves with weakly serrate margins. Infructescence axillary. Fruit to 3.2 cm diameter, the exocarp thin. Seeds one per fruit, to 2 cm diameter, with 10 weak longitudinal lines. Distribution: South America, also in Africa.

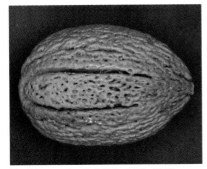

Humiriaceae - *Sacoglottis* (1.11×; 54.76 × 36.19)

Vantanea Aubl. Trees to 35 m tall. Leaves strongly coriaceous, the margins entire. Infructescence axillary or terminal. Fruit to 7.5 cm long, red, yellow, or brown when mature; the exocarp thick, carnose; the endocarp woody. Seeds one per fruit, to 6 cm long, with 5–8 longitudinal lines. Distribution: Costa Rica to Venezuela, Brazil, Peru, and Bolivia.

Humiriaceae - *Vantanea* (0.6×; 80 × 46.19)

ICACINACEAE

Trees, rarely shrubs or lianas. Leaves simple, alternate, entire. Fruit a drupe. Some genera aromatic.

Calatola Standley. Trees to 25 m tall. Infructescence axillary. Fruit a drupe, to 8 cm long, green or yellow when mature. Seeds one per fruit. Plants pubescent or glabrous. Distribution: Mexico to Chile.

Icacinaceae - *Calatola* (1.22×; 37.13 × 18.52)

Icacinaceae - *Calatola* (0.8×; 74.04 × 56.16)

Casimirella Hassler. Lianas or shrubs. Leaves with strongly reticulate venation and weakly decurrent bases, fragile when dry. Infructescence terminal. Fruit a drupe, to 7.5 cm diameter, green or yellow when mature, the mesocarp carnose and strongly aromatic. Seeds one per fruit. Distribution: Tropical America.

Icacinaceae - *Casimirella* (1.03x; 57.03 × 44)

Dendrobangia Rusby. Trees to 30 m tall. Leaves with fine tertiary venation imperceptible on the abaxial surface of the lamina. Infructescence axillary. Calyx persistent. Fruit a drupe, to 2 cm diameter, green or yellow when mature. Seeds one per fruit. Pubescence golden, short, and dense. Distribution: Panama to Peru and Bolivia.

Icacinaceae - *Dendrobangia* (2.03x; 19.48 × 11.25)

Discophora Miers. Trees to 10 m tall. Leaves with lightly decurrent venation. Infructescence axillary. Fruit a drupe to 1.5 cm long, convex or curved, black, with exposed white aril when mature. Seeds one per

Icacinaceae - *Discophora* (2.41x; 17.98 × 10.39)

fruit. Plants with short, dense, golden pubescence. Distribution: Panama to Peru and Bolivia.

Poraqueiba Aubl. Trees to 20 m tall. Leaves puberulent on abaxial surface, the tertiary venation conspicuous. Infructescence axillary. Fruit a drupe to 6 cm diameter, yellow or red when mature; the mesocarp thin with edible oil. Seeds one per fruit. At least one species widely cultivated for edible fruits. Distribution: Guianas, Brazil, to Peru.

Icacinaceae - *Poraqueiba* (8.1x; 7.62 × 5.46)

JUGLANDACEAE

Trees of tropical montane forest or temperate zones, included in this book because the seeds are commonly encountered on sandy beaches or floating in Amazonian rivers whose headwaters flow from the foothills of the Andes Mountains. Leaves compound, alternate, or opposite, with opposite and subopposite leaflets; the margins serrate. Infructescence axillary or terminal. Fruit a drupe. Plants pubescent and often aromatic; deciduous or caducifolious.

Juglans L. Trees to 25 m tall, restricted to forests above 1000 m elevation. Leaves imparipinnate and alternate. Fruit a drupe, to 4.5 cm diameter. Seed one per fruit. Tree used for lumber. Distribution: All of the Americas, Asia, and Europe, to 3500 m elevation.

Juglandaceae - *Juglans* (1.22x; 39.04 × 36.26)

JUNCACEAE

Terrestrial and aquatic grass-like herbs. Leaves simple, alternate, the base sheathing the stem. Infructescence terminal or axillary. Fruit a trivalved capsule. Seeds numerous per fruit.

Juncus L. Terrestrial or aquatic herbs. Leaves simple, alternate, narrow, sheathing the stem, closing at the apex. Infructescence terminal, arising from the closed, narrow leaf apex. Fruit a trivalved capsule, about 0.5 cm long, brown when mature, the stigmas sometimes persistent on the fruit. Seeds numerous per fruit, falling upon dehiscence. Plants often growing in Amazonian wetlands. Distribution: Worldwide.

Juncaceae - *Juncus* (19.79x; 3.21 × 2.45)

LACISTEMATACEAE

Family of only one genus, composed of trees. Leaves simple, alternate, entire. Stipules caducous, leaving a circular scar at each leaf node. Infructescence axillary. Fruit a capsule. Family segregated from the Flacourtiaceae.

Lacistema Sw. Trees to 25 m tall. Leaves with weakly serrate margins, especially on the upper half of the lamina. Fruit a capsule, bi- to trivalved, to 1 cm long, red when mature. Seeds 1–2 per fruit, surrounded completely by a white aril. Distribution: Mexico to Paraguay.

LAMIACEAE

Herbs or shrubs. Leaves simple, opposite or verticillate. Fruit a schizocarp with 1–4 nutlets when mature. Generally plants with strong fragrance or aroma. Stem square. In some texts still referred to as Labiatae.

Hyptis Jacq. Herbs or shrubs to 2 m tall. Leaves opposite, the margins dentate. Infructescence terminal or axillary. Fruit formed from 1–4 small nutlets, to 0.8 cm long, brown when mature and usually covered by

a persistent calyx. Pubescence short tomentose or glabrescent. Some species typical of disturbed areas or of wetlands in the Amazon region. Distribution: Southern United States and Mexico to Argentina, also in the Old World, to 2400 m.

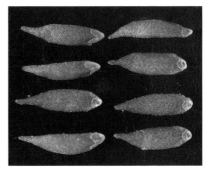

Lamiaceae - *Hyptis* (22.43x; 1.12 × 0.29)

LAURACEAE

Perhaps the most difficult family taxonomically, but easy to recognize to the family level. Mostly trees, sometimes shrubs, one genus of parasitic plants. Leaves simple, usually alternate, rarely opposite, strongly aromatic; the margins entire. Infructescence generally axillary. Fruit a drupe, with or without a persistent cupular calyx covering the base. Seeds one per fruit. Plants glabrous or pubescent with simple trichomes. The genera are typically separated on the basis of technical characteristics of floral morphology.

Aiouea Aubl. Trees to 28 m tall. Leaves sometimes subopposite. Infructescence axillary. Fruit to 2 cm long, black when mature; the mesocarp green; the subtending cupule red, to 3 x 1.1 cm. Plants glabrous or pubescent. Distribution: Mexico to Peru and southern Brazil.

Aniba Aubl. Trees or shrubs to 30 m tall. Leaves sometimes grouped at the branch apices, the abaxial leaf surface light brown or weakly golden in some species. Fruit to 3 cm long, black or bluish green

Lauraceae - *Aniba* (1.67x; 24.68 × 15.89)

when mature; the subtending cupule to 2.5 x 3 cm, sometimes woody, gray or brown when mature, larger than in other genera. Some species harvested for lumber. Distribution: Central America to Peru and Bolivia, to 2600 m elevation.

Beilschmiedia Nees. Trees to 30 m tall. Trunk with bark defoliating in round plates leaving circular orange-ochre scars. Leaves weakly glaucous abaxially. Infructescence borne from the stems. Fruit without a cupule, to 3 cm long, black when mature. Wood used in carpentry and construction. Distribution: Central America, Antilles, and Puerto Rico to Peru and Bolivia; also in the Paleotropics.

Lauraceae - *Endlicheria* (1.99x; 23.03 × 12.5)

Lauraceae - *Beilschmiedia* (1.57x; 28.11 × 12.64)

Cryptocarya R. Br. Trees to 30 m tall. Leaves rarely opposite. Fruit without cupule, to 2.5 cm diameter, yellow when mature. Distribution: Guianas to Peru, but more speciose in the Old World tropics.

Lauraceae - *Cryptocarya* (1.54x; 21.37 × 16.99)

Endlicheria Nees. Shrubs or trees to 30 m tall. Leaves sometimes verticellate with very reticulate and weakly decurrent venation in some species, the petioles longer than in most genera. Infructescence axillary or borne from the stems. Fruit to 3 cm long, black when mature, subtended by red cupule to 3.5 x 2.5 cm. Majority of species pubescent, the pubescence cream to golden, short, and very dense in some species. Distribution: Costa Rica to Paraguay, to 2400 m elevation.

Nectandra Rolander ex Rottb. Trees to 35 m tall. Leaves sometimes subopposite, the venation strongly to weakly decurrent, the base revolute toward the abaxial surface. Infructescence axillary or borne from the stems. Fruit to 2 cm long, black when mature, subtended by green cupule, to 1.5 x 1.4 cm; the mesocarp green. Many species with golden to brownish-red pubescence, short, very dense throughout. Wood harvested for lumber. Distribution: Southern United States to Argentina, to 2700 m elevation.

Lauraceae - *Nectandra* (3.53x; 14.94 × 11.59)

Lauraceae - *Nectandra* (2.79x; 15.45 × 8.37)

Ocotea Aubl. Shrubs or trees to 35 m tall. Leaves sometimes subopposite. Infructescence axillary or borne from the stems. Fruit to 2.5 cm long, sometimes round, black when mature, subtended by red cupule to 2.2 x 2.0 cm; the mesocarp green. Pubescence sometimes yellow-golden, dense throughout the plant in some species. Wood harvested for lumber. Distribution: Mexico to Uruguay, also in Africa, the Canary Islands, and Madagascar, to 2500 m elevation.

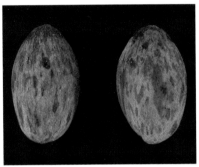

Lauraceae - *Ocotea* (2.58x; 15.49 × 9.39)

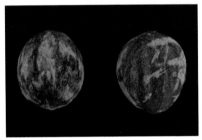

Lauraceae - *Ocotea* (2.05x; 13.4 × 11.7)

Rhodostemonodaphne Rohwer & Kubitzki. Trees to 20 m tall. Leaves sometimes subopposite. Fruit to 4.5 cm long, black when mature, subtended by red cupule, to 3 x 3.5 cm; the mesocarp green. Pubescence short, glaucous or golden on abaxial laminar surface and not very dense. Distribution: Costa Rica to Peru and Bolivia, to 1500 m elevation.

LECYTHIDACEAE

Trees, often large. Leaves simple, alternate, sometimes verticellate, sometimes grouped at the extreme of the branches; the margins finely serrate or crenate. Stipules caducous or absent. Infructescence axillary, terminal, or borne from the trunk. Fruit a pixis, in some genera the operculum is so small that the fruit is practically indehiscent, or in a few genera fruit a drupe or a berry. Seeds winged or arillate.

Bertholletia Humb. & Bonpl. Trees to 45 m tall, partially caducifolious in dry season. Infructescence

terminal. Fruit a pixis, practically indehiscent, to 16 cm diameter, brown when mature, the exocarp smooth and fleshy when mature; the mesocarp woody. Seeds up to 36 per fruit. Wood harvested for lumber. Widely cultivated and commercialized on a world market for edible seeds (i.e., Brazil nuts). Distribution: Guianas to Peru and Bolivia.

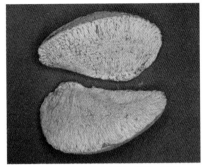

Lecythidaceae - *Bertholletia* (0.89x; 48.05 × 26.98)

Cariniana Casar. Trees to 45 m tall, caducifolious in dry season. Leaves with tertiary venation finely perpendicular to the principal glabrous or pubescent; the margins sometimes finely dentate-crenate. Infructescence terminal. Fruit a pixis, completely dehiscent, to 13 cm long, brown when mature; the exocarp woody, sometimes with dentate projections along the margin of the open operculum. Seeds winged, various per fruit. Wood harvested for lumber. Distribution: Brazil, Peru, and Bolivia.

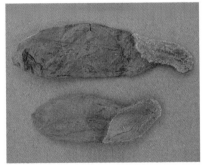

Lecythidaceae - *Cariniana* (1.21x; 51.4 × 15.09)

Couratari Aubl. Trees to 45 m tall, caducifolious in dry season. Infructescence axillary or terminal. Fruit a pixis, completely dehiscent, to 17 cm long, brown when dry; the exocarp woody. Seeds winged, various per fruit. Plants pubescent or glabrous. Wood harvested for lumber. Distribution: Costa Rica to Brazil, Peru, and Bolivia.

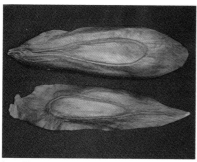

Lecythidaceae - *Couratari* (0.65x; 96.39 × 28.52)

Couroupita Aubl. Trees to 40 m tall. Leaves grouped at the extreme of the branches; the petioles and veins with short pubescence. Infructescence borne from the trunk. Fruit a pixis, indehiscent, to 18 cm diameter, brown when mature; the exocarp woody; the mesocarp white to yellow, fleshy, oxidizing to green upon contact with air, later to dark blue, usually with a fetid odor. Seeds many, embedded in the mesocarp. Distribution: Costa Rica to Peru.

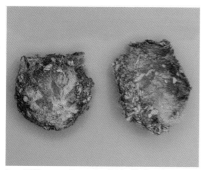

Lecythidaceae - *Couroupita* (2.01x; 15.63 × 12.63)

Eschweilera C. Mart. ex DC. Trees to 40 m tall. Leaves generally glabrous and coriaceous. Infructescence terminal or axillary. Fruit a pixis, to 8 cm diameter, brown when mature, completely dehiscent; the exocarp woody. Seeds 2–4 per fruit, with a rudimentary aril. Distribution: Mexico to Peru and Bolivia.

Lecythidaceae - *Eschweilera* (1.08x; 40.39 × 29.95)

Grias L. Trees to 10 m tall. Trunk usually without branches. Leaves usually large, grouped at the extremes of the branches, the margins entire. Infructescence borne from the trunk. Fruit a drupe to 8 cm long, green yellow when mature; the mesocarp carnose, edible. Seeds one per fruit. Cultivated for edible fruits. Distribution: Costa Rica to Brazil and Peru.

Lecythidaceae - *Grias* (1.37x; 43.33 × 20.93)

Gustavia L. Trees to 17 m tall. Leaves to 80 cm long, coriaceous, sometimes verticellate or grouped at the extremes of the branches; the margins serrate. Infructescence axillary or terminal. Fruit a pixis, to 10 cm long, brown when mature, completely dehiscent; the exocarp slightly woody. Seeds various per fruit. Plants often with a strongly fetid odor. Distribution: Costa Rica and Panama to Peru and Bolivia.

Lecythidaceae - *Gustavia* (1.38x; 20.23 × 19.31)

LOGANIACEAE

Lianas, shrubs, or trees. Leaves simple and opposite. Stipules present. Infructescence terminal or axillary. Fruit a capsule or berry. Seeds winged or not.

Potalia Aubl. Shrubs to 2 m tall, unbranched. Leaves to 60 cm long, grouped at the extreme of the single stem; the midvein triangular in cross section; the secondary veins inconspicuous to nearly invisible on the abaxial surface. Infructescence terminal, the peduncle orange. Fruit a berry to 1.7 cm long, green when mature, subtended by persistent calyx in the form of a cupule; the mesocarp fleshy, juicy. Seeds many per fruit. Distribution: Guianas to Peru.

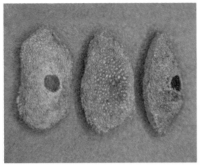

Loganiaceae - *Potalia* (7.04×; 4.93 × 2.69)

Strychnos L. Lianas, or sometimes shrubs. Leaves entire, sometimes coriaceous, trinerved from base. Infructescence terminal or axillary. Fruit a berry, to 5.5 cm, rarely to 12 cm diameter, yellow when mature; the exocarp subwoody; the mesocarp fleshy, juicy, transparent, orange, or yellow when mature. Seeds 1–8, rarely up to 25 per fruit. Some species with straight or weakly curved axillary spines. Plants usually glabrous, rarely pubescent. Distribution: Central America to Peru, also in the Old World tropics.

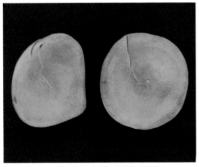

Loganiaceae - *Strychnos* (1.28×; 25.04 × 23.8)

Loganiaceae - *Strychnos* (1.49×; 18.37 × 14.61)

Loganiaceae - *Strychnos* (2.71×; 20.51 × 16.52)

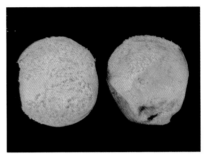

Loganiaceae - *Strychnos* (3.62×; 8.33 × 7.92)

LORANTHACEAE

Parasitic, hemiepiphytic herbs, shrubs, or lianas. Leaves simple and opposite, rarely alternate, usually succulent, the venation rather inconspicuous. Fruit a berry or a drupe.

Psittacanthus Mart. Epiphytic parasite. Leaves weakly asymmetrical, alternate in some species, the midvein and secondary veins rather inconspicuous, the tertiary venation imperceptible. Infructescence axillary or terminal. Fruit a drupe to 3 cm long, black when mature, subtended by prominent receptacle forming a cupule, sometimes reddish; the ostiole very deep; the mesocarp green, sticky. Seeds one per fruit. Distribution: Colombia, Peru, Brazil, and Bolivia.

Loranthaceae - *Psittacanthus* (2.24×; 20.47 × 9.25)

LYTHRACEAE

Herbs, shrubs, or trees. Leaves simple, opposite. Plants sometimes with inconspicuous stipules. Infructescence axillary or terminal. Fruit a capsule. Seeds short-winged.

Cuphea P. Browne. Herbs or shrubs to 1 m tall, some species lianescent. Plants with short pubescence, sometimes asperous. Leaves opposite, entire, variable in size. Infructescence terminal or axillary. Fruit a loculicidal capsule, to about 4 cm long, partly or completely surrounded by persistent calyx, brown when mature. Seeds many per fruit, short-winged. Some species cultivated as ornamentals. Leaves and flowers medicinal. Common in disturbed areas. Distribution: Mexico to Argentina to 4000 m elevation.

Lafoensia Vand. Trees to 25 m tall. Leaves opposite, more or less coriaceous, with finely parallel secondary venation and a marginal collecting vein. Infructescence terminal. Fruit a loculicidal capsule, to 7 cm diameter, brown when mature, subtended by persistent calyx. Seeds winged, many per fruit. Distribution: Mexico to Peru and Bolivia.

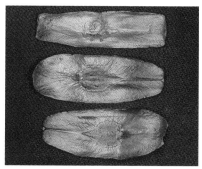

Lythraceae - *Lafoensia* (1.28x; 37.44 × 13.14)

Physocalymma Pohl. Trees to 30 m tall. Leaves asperous, caducous. Infructescence axillary. Fruit a capsule, to 1 cm diameter, dark purple to purple-black when mature, subtended by persistent calyx. Seeds many per fruit. Distribution: South America, to 1200 m elevation.

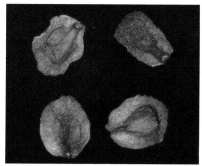

Lythraceae - *Physocalymma* (5.93x; 3.59 × 3.07)

MAGNOLIACEAE

Trees, restricted primarily to montane forest, with few species in the Amazon. Leaves simple, alternate, sometimes caducous. Stipule terminal, very conspicuous, caducous, leaving a circular scar around the stem at each node. Infructescence axillary or terminal. Fruit an aggregate, originating from an apocarpous gynoecium. Seeds many per fruit, covered with a red aril when mature. Plants glabrous or pubescent.

Magnolia L. Trees to 10 m tall, caducifolious. Infructescence axillary or terminal. Fruit an aggregate. Widely cultivated as ornamentals. Distribution: Temperate regions of North America and Asia, to 3600 m elevation.

Magnoliaceae - *Magnolia* (2.62x; 10.2 × 6.61)

Talauma Juss. Trees to 20 m tall, weakly aromatic. Leaves with caniculate petiole. Infructescence terminal. Fruit to 11 cm long, the exocarp brown, woody, opening irregularly to expose the seeds when mature. Seeds arillate, reddish-orange when mature, many per fruit. Distribution: Tropical America, Oceania, and principally in Asia.

Magnoliaceae - *Talauma* (2.31x; 10.46 × 8.07)

MALPIGHIACEAE

Shrubs, trees, and lianas. Leave simple, opposite, rarely verticellate; the margins entire. Stipules axillary, often intrapetiolar, positioned between the stem and base of petiole. Infructescence terminal or axillary. Fruit a drupe or samara. Plants usually with glands on the leaf or the petiole, and on the calyx in flower and fruit. Pubescence of T-shaped, malpighiaceous trichomes or hairs, rarely stellate or simple.

Banisteriopsis C. B. Robinson. Lianas. Infructescence axillary, terminal, or borne from the stems, with conspicuous bracts. Glands present or absent on the leaf or petiole. Fruit a samara, united in groups of 1–3, with the apical wing to 4 cm long, red when immature, brown when mature. Plants glabrous or pubescent; sometimes the abaxial laminar surface glaucous to golden pubescent. One species known as *Ayahuasca* is famous in the Amazon for its hallucinogenic properties. Distribution: Mexico and Guianas to Peru and Argentina.

Malpighiaceae - *Banisteriopsis* (1.57x; 37.7 × 12.7)

Bunchosia Rich. ex Kunth. Trees to 7 m tall. Stipules inconspicuous. Leaves usually with a pair of glands on the base of the midvein on the abaxial surface, or glands sometimes absent. Infructescence axillary. Fruit a berry, to 4 cm diameter, red when mature, subtended by persistent calyx; the mesocarp red, smooth. Seeds 1–2 per fruit. Many species with thick pubescence of T-shaped hairs. Some species cultivated for their edible fruits. Distribution: Mexico and West Indies to Peru and Bolivia.

Malpighiaceae - *Bunchosia* (1.88x; 23.65 × 15.67)

Malpighiaceae - *Bunchosia* (1.61x; 29.67 × 18.22)

Byrsonima Rich. ex Kunth. Shrubs or trees to 30 m tall. Leaves often with anastomosing secondary venation. Stipules axillary, usually intrapetiolar, positioned between the stem and base of petiole. Infructescence terminal. Fruit a drupe, to 1.5 cm diameter, yellow, orange, red, or black when mature, subtended by persistent, glandular calyx. Seeds one per fruit. Plants typically with dense pubescence of simple or T-shaped hairs, golden or reddish in color, or rarely glabrescent. The bark of various species is rich in tannins. In rare cases some tree species are used for lumber. Distribution: Southern United States, Central America to Peru and Argentina, to 1600 m elevation.

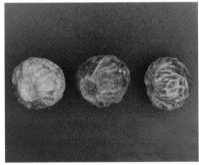

Malpighiaceae - *Byrsonima*. (2.67x; 7.06 × 6.59)

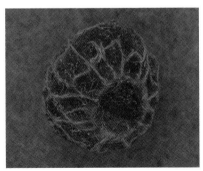

Malpighiaceae - *Byrsonima* (5.78x; 7.53 × 6.72)

Dicella Griseb. Lianas. Leaves with finely reticulate secondary venation; glands inconspicuous on the abaxial laminar surface, or absent. Stipules small, inconspicuous. Fruit a dry nut surrounded by five nerved wings formed by the sepals, the wings to 5 cm long, unequal in length. Distribution: Tropical South America.

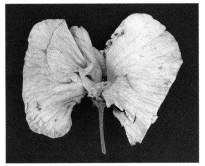

Malpighiaceae - *Hiraea* (1.54x; 35.21 × 22.4)

Malpighiaceae - *Dicella* (1.03x; 60.21 × 59.9)

Heteropterys H.B.K. Lianas. Leaves with a pair of glands on the petiole at the base of the lamina, adaxially, or glands absent. Stipules very small or absent. Infructescence terminal or axillary, with conspicuous bracts or bracteoles. Fruit a samara, in clusters of 1–3, the apical wing to 5 cm long, brown when mature; subtended by persistent calyx and stamens; calyx glands present or absent.

Mascagnia (Bert. ex DC.) Colla. Lianas. Leaves with glands on the base of the lamina or on the petiole. Infructescence axillary, terminal, or borne from the stems with bracteoles in some species. Fruit a samara, united in clusters of three, the wing concentric, entire or divided in two, to 7 cm diameter, generally with additional smaller wings, white or brown when mature; subtended by persistent, glandular calyx. Pubescence simple or compound; some species with completely glaucous abaxial laminar surface. Distribution: Mexico and Guianas to Peru, Paraguay, and Argentina, to 700 m elevation.

Malpighiaceae - *Mascagnia* (1.43x; 36.59 × 33.41)

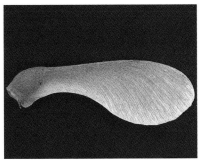

Malpighiaceae - *Heteropterys* (3.7x; 17.31 × 5.8)

Hiraea Jacq. Lianas. Leaves with a pair of stipules at the apical end of the petiole; glands present on the laminar base or the petiole apex. Infructescence axillary or borne from the stems with bracteoles. Fruit a samara, united in clusters of three, to 4 cm diameter, the wing subconcentric or auricular, generally with other reduced wings, white or brown when mature; subtended by persistent, glandular calyx. Plants glabrous or pubescent. Distribution: Mexico and Central America to Argentina.

Malpighiaceae - *Mascagnia* (0.86x; 58.15 × 65.56)

Tetrapterys Cav. Lianas. Infructescence terminal or axillary, with bracteoles. Fruit a samara, wings four, the largest wing to 4 cm long, sometimes additional smaller wings present, brown when mature; subtended by persistent, glandular calyx. Plants pubescent or glabrous, sometimes with inconspicuous glands. Distribution: Mexico to Brazil, Peru and Bolivia.

Malpighiaceae - *Tetrapterys* (1.24x; 39.66 × 7.35)

Malpighiaceae - *Tetrapterys* (1.6x; 30.33 × 9.07)

MALVACEAE

Herbs and shrubs, rarely small trees. Leaves alternate, simple, entire, or palmately lobed, often trinerved from base of lamina; the petiole with swollen pulvinus at base and apex. Stipules persistent or caducous. Infructescence axillary or terminal. Fruit a loculicidal capsule, rarely a berry. Generally pubescent, the hairs stellate or simple.

Hibiscus L. Herbs or shrubs to 2 m tall. Leaves entire or palmately lobed, the margins serrate, crenate, or dentate, the adaxial laminar surface lightly asperous. Infructescence axillary or terminal. Fruit a 5-valved capsule, to 2.5 cm long, brown when mature, subtended to two-thirds length by persistent calyx. Seeds many per fruit. Plants with pubescence of very short stellate hairs throughout, sometimes simple on the fruit. Many species cultivated as ornamentals. Distribution: Southern United States, Mexico, and Cuba to Bolivia and Argentina, to 3500 m elevation, also in Asia.

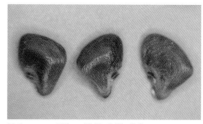

Malvaceae - *Hibiscus* (6.14x; 3.52 × 2.6)

Pavonia Cav. Shrubs to 1.5 m tall. Leaves usually entire or weakly lobed; the margins dentate, crenate, or serrate; the base conspicuously asymmetric in many species. Stipules axillary, conspicuous. Infructescence terminal. Fruit a 5-valved capsule, to 1.5 cm long, brown when mature, subtended and covered by persistent calyx; the five capsular valves terminated by a spine. Plants with pubescence of stellate hairs throughout. Distribution: Southern United States, Mexico, and West Indies to Peru and Paraguay, also in the Old World tropics.

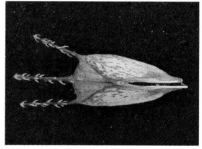

Malvaceae - *Pavonia* (6.39x; 9.04 × 2.7)

Sida L. Shrubs to 1 m tall. Leaves with dentate or serrate margins. Infructescence axillary, terminated by conspicuous persistent stigmas of the clustered fruits, collectively subtended by persistent calyx. Fruit a 5-valved capsule, to 1 cm long, with persistent stigmas, dry and brown when mature. Seeds many per fruit. Plants with dense pubescence of small stellate hairs throughout. Many species common in disturbed areas. Distribution: Tropics and subtropics of the world.

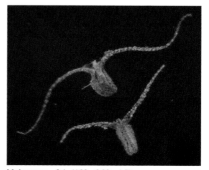

Malvaceae - *Sida* (4.93x; 2.38 × 1.5)

MARANTACEAE

Herbs, rarely lianas. Plants rhizomatous. Leaves simple, alternate, broad, and entire, sometimes variegated or colored, the lamina exposing conspicuous fine hairs or fibers when ripped. Petiole apex with a swollen pulvinus at the junction with the lamina, permitting movement of the leaf. Fruit a capsule, surrounded by multiple bracts that are often large, colorful, and attractive.

Calathea G. Mey. Herbs, often acaulescent. Leaves to 60 cm long, the abaxial surface reddish to dark purple or with white or reddish lines; the secondary and tertiary veins finely parallel, differentiated on the adaxial laminar surface, the petiole to 80 cm long. Fruit a 3-valved capsule, to 2 cm long, yellow or orange when mature, subtended by and covered by persistent bracts to 5 cm long. Seeds 1–3 per fruit, blue when mature, covered at the base by a white aril. Many species are common in open sunny areas. Distribution: Mexico to Argentina.

Marantaceae - *Calathea* (3.39x; 7.81 × 3.75)

Hylaeanthe A. M. E. Jonker & Jonker. Herbs to 0.5 m tall, rhizomatous. Plants annual, the leaves present just slightly before flowering and fruiting. Leaves to 60 cm long, weakly succulent, the venation very fine and parallel. Infructescence terminal. Fruit a capsule. Seeds various per fruit. Distribution: South America.

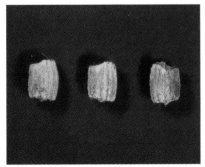

Marantaceae - *Hylaeanthe* (3.07x; 4.83 × 3.45)

Ischnosiphon Koern. Climbing herbs. Leaves entire, the venation parallel, very lightly distinguished on the adaxial laminar surface; the petioles with a winged section near base and sheathing the stem. Plants often with smooth pubescence. Infructescence axillary or terminal. Fruit a drupe. Seeds various per fruit. Distribution: Panama to Peru and Bolivia.

Marantaceae - *Ischnosiphon* (1.8x; 31.12 × 5.12)

Monotagma K. Schum. Herbs with very short stems or acaulescent. Leaves to 60 cm long, the venation strongly parallel; the petiole to 40 cm long and winged; the secondary and tertiary veins differentiated only on the adaxial laminar surface; the midvein terminating before the laminar apex. Fruit a capsule, to 2 cm long, brown when mature, subtended and covered by a bract to 5 cm. Seeds one per fruit. Distribution: Venezuela to Peru.

Marantaceae - *Monotagma* (2.67x; 15.6 × 3.97)

MARCGRAVIACEAE

Hemiepiphytes and lianas. Leaves simple, alternate, entire. Leaves sessile or subsessile, with very short petioles. Stipules persistent. Infructescence terminal with large nectaries. Fruit a capsule.

Marcgravia L. Hemiepiphytes. Leaves coriaceous, sessile or subsessile, typically with blackish glandular punctations along the margin; the venation inconspicuous. Infructescence composed of approximately 15 fruits in a concentric circle with a persistent saccate nectary in the center. Fruit a 5- to 10-valved capsule, to 1.5 cm diameter, brown when mature, subtended by persistent calyx. Seeds many per fruit. Distribution: Central and South America, and the Caribbean as far north as Cuba.

Norantea Aubl. Lianas. Leaves entire, succulent, the adaxial surface glossy, the secondary venation inconspicuous. Infructescence terminal with persistent, showy floral bracts. Fruit a capsule, red or brown when mature. Seeds various per fruit. Distribution: Belize, Honduras, and Jamaica to Brazil and Peru.

Marcgraviaceae - *Norantea* (4.28×; 4.7 × 1.61)

Souroubea Aubl. Hemiepiphytes. Leaves succulent, the secondary venation weakly conspicuous, the tertiary venation inconspicuous. Infructescence a spike with persistent saccate nectaries. Fruit a 5-valved capsule, to 1.7 cm long, green or brown when mature, subtended by persistent calyx. Seeds various per fruit, with orange arils. Distribution: Belize to Guianas, Brazil, and Peru.

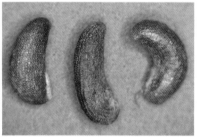

Marcgraviaceae - *Souroubea* (8.1×; 4.17 × 1.74)

MAYACACEAE

Aquatic or terrestrial herbs, sometimes submerged. Stems sometimes branched and matted, sprawling or erect. Leaves simple, alternate, linear or lanceolate, and sessile. Infructescence terminal. Fruit a loculicidal capsule.

Mayaca Aubl. Aquatic or terrestrial herbs, sometimes submerged. Stems single or branched and matted, sprawling or erect. Leaves simple, alternate, linear or lanceolate, and sessile. Infructescence terminal. Fruit a loculicidal capsule subtended by per-

sistent calyx and peduncle, about 0.2–0.3 cm long, brown when mature. Distribution: Southern United States to Paraguay, one species in Africa; often growing in wetland ecosystems at edges of floating islands of aquatic vegetation.

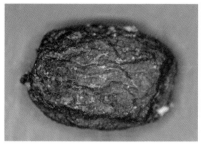

Mayacaceae - *Mayaca* (13.54×; 4.04 × 2.69)

MELASTOMATACEAE

Composed principally of shrubs, but also herbs, trees, lianas, and some hemiepiphytes. Leaves simple, opposite, sometimes verticellate, entire or lightly serrate; the secondary venation trinerved from base, or sometimes 5-nerved; the tertiary venation conspicuously parallel and perpendicular to the secondary veins and midvein. Infructescence terminal or axillary. Fruit a berry in species with inferior ovaries, or a loculicidal capsule in species with superior ovaries. Seeds always very small, often microscopic, and many per fruit. Confused only with Loganiaceae when sterile, but the leaves of Loganiaceae are more coriaceous and glabrous. Plants usually with some form of pubescence, especially on the leaves, rarely glabrous. Some genera and species are characterized by anisophylly (one leaf larger than the other at a node) or by the presence of ant domatia.

Adelobotrys DC. Lianas. Infructescence terminal. Fruit a capsule opening in various parts, to 1 cm long, brown when mature. Seeds many per fruit, without arils, elongate, apparently dispersed by wind. Plants usually pubescent throughout. Distribution: Mexico to Bolivia.

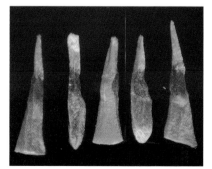

Melastomataceae - *Adelobotrys* (23.17×; 1.74 × 0.46)

Bellucia Raf. Trees to 20 m tall. Leaves typically large. Infructescence borne from the stems or trunk. Fruit a berry, to 3 cm diameter, yellow and edible when mature. Seeds tiny, many per fruit. Plants generally growing in disturbed areas. Distribution: Southern Mexico to Bolivia, to 1500 m elevation.

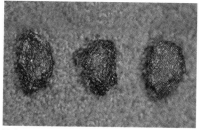

Melastomataceae - *Bellucia* (24.02x; 0.81 × 0.5)

Clidemia D. Don. Shrubs to 3 m tall. Leaves anisophyllous in some species; the petioles of some species with domatia. Infructescence axillary, sometimes borne from the trunk. Fruit a berry, usually pubescent, to 1 cm diameter, blue, violet, or black when mature, terminated by persistent stigma. Seeds tiny, many per fruit. Plants typically with pubescence of stellate hairs throughout. Distribution: Southern Mexico to Bolivia.

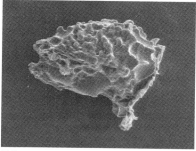

Melastomataceae - *Clidemia* (96.3x; 0.49 × 0.3)

Henriettella Naudin. Shrubs or trees to 10 m tall. Infructescence axillary or borne from the stems or trunk. Fruit a berry, flattened on upper surface, to 1

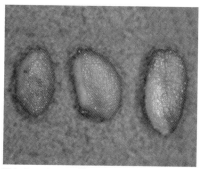

Melastomataceae - *Henriettella* (20x; 1.44 × 0.7)

cm diameter, black when mature. Seeds tiny, many per fruit. Plants glabrous or pubescent. Distribution: Central America to Bolivia.

Graffenrieda DC. Shrubs or small trees to 15 m tall, sometimes lianas. Leaves entire, often large and coriaceous, 3-veined from the base with anastomosing marginal veins, the lamina and petiole glabrous or with fine reddish-brown tomentose pubescence. Infructescence terminal. Fruit a capsule, to 0.7 cm diameter, prominently ribbed. Seeds pyramidal, elongate, dry, many per fruit, dispersed by wind. Distribution: Central America and the West Indies to Bolivia and Brazil, to 1500 m elevation, often growing in wetland habitats.

Leandra Raddi. Shrubs to 3 m tall. Leaf margins dentate. Infructescence terminal. Fruit a berry, sessile, to 1 cm diameter, black when mature. Seeds tiny, many per fruit. Plants usually pubescent throughout. Easily confused with *Miconia* and *Ossaea*, technically differing in characteristics of the flower petals. Distribution: Belize and Guatemala to Bolivia.

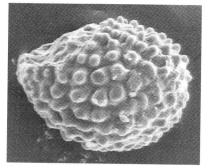

Melastomataceae - *Leandra* (183.07x; 0.33 × 0.25)

Loreya DC. Trees to 10 m tall. Leaves usually large, sometimes verticillate. Infructescence borne from the stems or trunk. Fruit a berry, yellow or purple when mature. Seeds tiny, many per fruit. Plants usually pubescent, often densely covered by conspicuous, showy red hairs. Distribution: Nicaragua to Peru and Bolivia, to 1400 m elevation.

Melastomataceae - *Loreya* (34.39x; 0.77 × 0.37)

Maieta Aubl. Shrubs to 2 m tall. Leaves anisophilous; domatia present at the base of the adaxial laminar surface, but only on the largest leaf at each node. Infructescence axillary. Fruit a berry to 1 cm diameter, pubescent, dark purple when mature. Seeds tiny, many per fruit. Plants usually somewhat pubescent throughout. Distribution: Guianas to Bolivia.

Miconia R. & P. Shrubs or trees to 20 m tall. Leaves morphologically variable but always with strongly parallel tertiary venation perpendicular to the central and secondary veins. Infructescence axillary. Fruit a berry to 2 cm diameter, generally black, sometimes gray or blue when mature, often pubescent, and sometimes subtended by persistent calyx. Seeds three to many per fruit. Plants glabrous or with dense pubescence of stellate hairs. One of the most diverse genera in the Neotropics. Distribution: Southern Mexico to Bolivia and Paraguay, to 3700 m elevation.

Melastomataceae - *Miconia* (28.68x; 0.75 × 0.65)

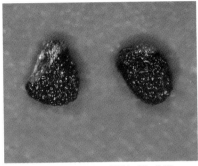

Melastomataceae - *Miconia* (59.26x; 0.8 × 0.63)

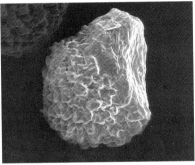

Melastomataceae - *Miconia* (16.72x; 1.44 × 0.89)

Ossaea DC. Shrubs to 4 m tall. Leaves sometimes anisophyllous, the lamina very thin. Infructescence axillary and terminal. Fruit a berry, to 1 cm diameter, violet or dark purple when mature. Seeds tiny, many per fruit. Plants usually pubescent throughout. Distribution: Mexico to Peru and Bolivia.

Melastomataceae - *Miconia* (24.34x; 0.71 × 0.6)

Melastomataceae - *Ossaea* (144.97x; 0.34 × 0.25)

Salpinga Mart. ex DC. Herbs to 0.5 m tall. Leaves nearly transparent when pressed and dried. Infructescence axillary, scorpioid. Fruit a capsule, to 2 cm long, brown when mature. Seeds tiny, free, many per fruit. Distribution: Peru.

Melastomataceae - *Miconia* (27.94x; 1.05 × 1.03)

Melastomataceae - *Salpinga* (60.53x; 0.8 × 0.68)

Tibouchina Aubl. Herbs in lowland Amazonia forest, shrubs or small trees in montane forests. Leaves anisophyllous. Infructescence axillary or terminal. Fruit a capsule, to 2 cm long, brown when mature. Seeds tiny, free, many per fruit. Plants glabrous or pubescent throughout. Distribution: Southern Mexico to Bolivia.

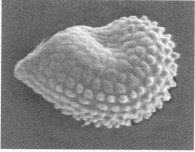

Melastomataceae - *Tibouchina* (193.65x; 0.29 × 0.19)

Tococa Aubl. Shrubs to 4 m tall. Leaves with weakly dentate margins, the petioles with domatia at apex. Infructescence axillary or terminal. Fruit a berry, pubescent, to 1 cm diameter, dark purple or black when mature, subtended by persistent calyx. Seeds tiny, many per fruit. Plants usually somewhat pubescent throughout. Distribution: Southern Mexico to Peru and Bolivia, to 1500 m elevation.

Melastomataceae - *Tococa* (21.8x; 1.61 × 0.85)

MELIACEAE

Trees, rarely shrubs. Leaves alternate, compound, imparipinnate or paripinnate. Fruit a loculicidal or septicidal capsule. Seeds winged or arillate. Family important for the quality and value of wood, especially as the source of the prized mahogany (*Swietenia*)and Spanish cedar (*Cedrela*). Bark of *Cedrela, Guarea,* and *Trichilia* produce tannins utilized on a small scale to manufacture insecticides, medications, and dyes.

Cabralea Adr. Juss. Trees to 25 m tall. Leaves paripinnate, clustered at branch apices, the leaflets opposite or subopposite, conspicuously curved with an asymmetric base, very fragile when dry. Infructescence axillary or borne from the stems or trunk. Fruit a loculicidal capsule, 4- to 5-valved, to 4 cm diameter, brown when mature. Seeds 1–2 per locule, covered with red aril. Wood used in carpentry and construction. Distribution: Costa Rica to Argentina, to 2500 m elevation.

Meliaceae - *Cabralea* (2.56x; 12.4 × 9.59)

Cedrela P. Browne. Trees to 40 m tall. Leaves paripinnate, the leaflets opposite or subopposite with asymmetric base. Infructescence terminal. Fruit a septicidal capsule dehiscing from the apex in five valves, to 8 cm long, brown with tan spots when mature; the central column subwoody. Seeds winged,

Meliaceae - *Cedrela* (1.4x; 41.44 × 15.68)

many per fruit. Plants glabrous or pubescent, generally caducifolious. Wood highly valued for use in fine carpentry and furniture, construction, and to make boats (known as Spanish cedar or cedro). Widely cultivated in forestry plantations. Leaves, resin, flowers, and bark medicinal. Distribution: Mexico and West Indies to Argentina, to 3600 m elevation.

Guarea Allam. ex L. Trees to 30 m tall; some species shrubs. Leaves paripinnate with an indeterminate apical bud, the leaflets opposite. Infructescence axillary or borne from the stems. Fruit a loculicidal capsule, 4- to 6-valved, to 8 cm diameter, brown or reddish when mature; smooth or tuberculate; the exocarp subwoody; the mesocarp white, mealy. Seeds 1–2 per locule, covered with red aril, . Plants glabrous or pubescent. Wood used in furniture and construction. Shoots, bark, and fruits reportedly medicinal. Distribution: Mexico and West Indies to Argentina, also in Africa, to 3000 m elevation.

Meliaceae - *Guarea* (2.2x; 13.89 × 8.37)

Meliaceae - *Guarea* (3.22x; 11.28 × 7.96)

Swietenia Jacq. Trees to 40 m tall. Leaves paripinnate, the leaflets opposite or subopposite with asymmetric bases. Infructescence axillary. Fruit a septicidal capsule, dehiscing from the apex in five valves, to 15 cm long; the central column subwoody, brown when mature; the exocarp woody. Seeds winged, many per fruit. Plants caducifolious in dry season. Wood high quality and valuable, used in fine furniture and carpentry (known as mahogany or caoba). Widely cultivated in forestry plantations. Distribu-

tion: Mexico to West Indies to Argentina, to 3600 m elevation.

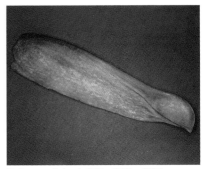

Meliaceae - *Swietenia* (0.62x; 107.01 × 27.35)

Trichilia P. Browne. Trees to 35 m tall. Leaves imparipinnate, the leaflets opposite and alternate, smaller toward the base of the leaf; generally with a pair of extremely reduced, stipule-like leaflets near the base of the petiole. Infructescence axillary. Fruit a loculicidal capsule, bi- to trivalved, to 3.5 cm long, smooth or tuberculate, brown, green, orange, or yellow when mature. Seeds 1–2 per fruit, with tan, orange, or red aril. Plants glabrous or pubescent. Wood used in general carpentry. Distribution: Mexico, Cuba, and Puerto Rico to Argentina and Paraguay, also in Africa and Malesia, to 3000 m elevation.

Meliaceae - *Trichilia* (6.03x; 7.98 × 3.23)

Meliaceae - *Trichilia* (5.03x; 6.46 × 5.07)

MEMECYLACEAE

Usually trees, rarely shrubs. Leaves simple, opposite, entire, typically glabrous; the venation pinnate, sometimes inconspicuous. Infructescence axillary or borne from the trunk. Fruit a berry. Seeds 1–5 per fruit. Family segregated from the Melastomataceae. Easily confused in the herbarium and the field with the Myrtaceae, but lacking translucent punctations and aromatic odor of that family.

Mouriri Aubl. Shrubs or trees to 20 m tall. Leaves of some species with inconspicuous secondary and tertiary leaf venation; other species with sessile leaves. Infructescence axillary or borne from the trunk. Fruit a berry, to 1 cm diameter, yellow, orange, or red when mature. Distribution: Mexico to Bolivia.

Memecylaceae - *Mouriri* (3.26x; 11.1 × 8.18)

Memecylaceae - *Mouriri* (2.5x; 10.64 × 7.08)

MENISPERMACEAE

Primarily lianas, rarely shrubs. Leaves simple, alternate, generally entire, some genera with elongated petiolate pulvinus; the secondary venation sometimes 3- to 5-nerved. Infructescence axillary or borne from the trunk. Fruit a drupe. Seeds very distinct. Species used as medicine, sweeteners, contraceptives, and arrow poisons.

Abuta Aubl. Lianas, but sometimes growing in early stages as shrubs. Leaves 3- to 5-nerved, the petioles with pulvinus present at both extremes, flexed to-

ward the apex. Infructescence axillary. Fruit a drupe, to 4 cm long, yellow or orange when mature. Seed one per fruit. Plants glabrous or pubescent. Distribution: Mexico to Bolivia.

Menispermaceae - *Abuta* (2.07x; 23.11 × 11.33)

Anomospermum Miers. Lianas. Leaves 3- to 5-nerved, coriaceous, the petioles elongate and pulvinate. Fruit a drupe, sometimes in pairs from an apocarpous gynoecium, to 5 cm long, orange when mature. Seed one per fruit. Distribution: Brazil and Peru.

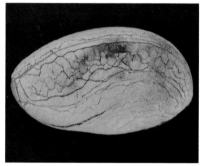

Menispermaceae - *Anomospermum* (1.74x; 36.1 × 20.79)

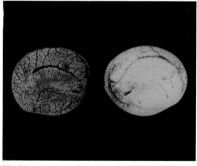

Menispermaceae - *Anomospermum* (1.28x; 22.73 × 18.76)

Borismene Barneby. Lianas. Leaves trinerved, the petiole pulvinate. Fruit a drupe, solitary or in pairs from an apocarpous gynoecium, to 2.5 cm long, red, the mesocarp orange when mature; seeds one per fruit. Easily confused vegetatively with *Disciphania* and *Odontocarya*. Distribution: Brazil, Ecuador, and Peru.

Menispermaceae - *Borismene* (1.9×; 19.11 × 13.39)

Chondodendron R.& P. Lianas. Leaves palmately veined, glaucous, the petioles elongate and weakly pulvinate, curved toward the apex. Infructescence axillary. Fruit a drupe, in clusters of three per pedicel from an apocarpous gynoecium, to 1.5 cm long, yellow to black when mature, each fruit constricted at the base giving the impression of being short stipitate. Seed one per fruit. Distribution: Panama to Bolivia.

Cissampelos L. Lianas or climbing herbs. Leaves somewhat peltate with a characteristic apicule at the apex of the principal nerve; the petioles elongate; the secondary venation palmate. Fruit a drupe, lightly flattened, to 1.0 cm long, red or orange when mature. Seeds one per fruit. Distribution: Tropical America and the Old World.

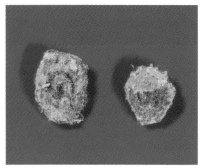

Menispermaceae - *Cissampelos* (4.08×; 6.06 × 4.56)

Curarea Barneby & Krukoff. Lianas. Leaves palmately veined, glaucous or with tomentose pubescence; the petioles elongate and weakly pulvinate. Infructescence borne from the trunk, sometimes up to 1 m long. Fruit a drupe to 1.5 cm long, in clusters from an apocarpous gynoecium; the drupes rugose, with dense pubescence, orange when mature. Seeds one per fruit. Distribution: Brazil, Ecuador, and Peru.

Menispermaceae - *Curarea* (2.75×; 18 × 7.73)

Disciphania Eichler. Leaves palmately veined, the petioles elongate and pulvinate. Infructescence spicate, elongate, sometimes arising from the base of the stems near the soil. Fruit a drupe, to 1.5 cm long, black when mature, clustered from an apocarpous gynoecium; the mesocarp mucilaginous. Seeds one per fruit. Plants glabrous or pubescent. Distribution: Venezuela to Brazil, Peru, and Bolivia.

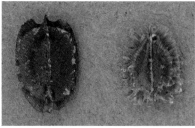

Menispermaceae - *Disciphania* (2.73×; 12.02 × 7.87)

Odontocarya Miers. Lianas. Leaves cordate, the secondary venation palmate, the petioles thin and curved at base. Infructescence axillary or borne from the trunk. Fruit a drupe, to 1.5 cm long, yellow when mature, arranged in groups of three from an apocarpous gynoecium; the mesocarp transparent and sticky. Seeds one per fruit. Distribution: Southern Mexico to Bolivia.

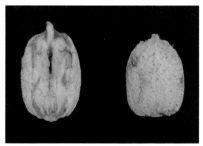

Menispermaceae - *Odontocarya* (2.5×; 13.43 × 8.31)

Sciadotenia Miers. Lianas. Leaves 3- to 5-nerved. Fruit a drupe to 1 cm long, orange when mature, clustered in groups of 3–5 per pedicel from an apocarpous gynoecium. Seeds one per fruit. Distribution: Brazil, Peru, and Bolivia.

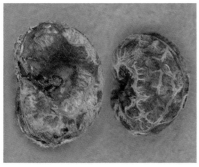

Menispermaceae - *Sciadotenia* (2.84×; 15.11 × 10.78)

MONIMIACEAE

Family composed aromatic shrubs and trees. Leaves simple, opposite. Infructescence axillary. Fruit apocarpous.

Mollinedia R. and P. Shrubs or trees to 15 m tall. Leaves with entire or weakly serrate margins on the apical one-third of the lamina. Infructescence axillary, terminal, or rarely borne from the trunk. Fruits apocarpous, composed of 6–25 monocarps attached to a subtending red receptacle; each monocarp to 3 cm long, sessile or stipitate, purple to black when ma-

Monimiaceae - *Mollinedia* (2.62×; 18.23 × 10.44)

Monimiaceae - *Mollinedia* (3.15×; 14.3 × 7.82)

ture; the mesocarp yellow or orange, very thin. Seeds one per monocarp. Plants glabrous or pubescent. Distribution: Tropical or subtropical, to 2600 m elevation.

MORACEAE

One of the most diverse families in the tropics of the world, composed primarily of trees, but also with herbs, shrubs, and hemiepiphytes. Leaves simple, alternate. Stipules terminal, conspicuous, leaving a scar in the form of a ring at each node. Latex abundant, usually white or cream; in some species yellow or watery brown. Infructescence axillary. Fruit a drupe, syconium, or multiple formed by many fused monocarps.

Batocarpus H. Karst. Trees to 30 m tall. Trunks conspicuously reddish in some species. Leaves usually coriaceous, somewhat scabrous, the margins conspicuously serrate. Terminal stipule not forming a conspicuous circular scar when falling. Latex white, abundant. Infructescence axillary. Fruit a multiple to 6.5 cm diameter, green or yellow when mature. Seeds various per multiple fruit. Distribution: Tropical America.

Moraceae - *Batocarpus* (1.6×; 18.61 × 10.46)

Brosimum Sw. Trees to 40 m tall. Leaves entire, scabrous in some species. Terminal stipules caducous, generally leaving a circular scar. Latex white. Infructescence axillary, technically developing into a multiple fruit with two or more monocarps. Fruit a multiple to 4 cm diameter, completely round or flat at the

Moraceae - *Brosimum* (1.92×; 16.35 × 14.81)

apex, often terminated by persistent stigma; some-times asperous due to the presence of scales; yellow, orange, green, or purple when mature; the mesocarp yellow or orange. Seeds 1–3 per multiple fruit. Plants glabrous or pubescent. Some species used for wood, others for edible fruit and latex. Distribution: Mexico, Costa Rica, and Guianas, to Brazil and Peru.

Moraceae - Brosimum (1.38×; 16 × 13)

Castilla Cervantes. Trees to 30 m tall. Leaves cadu-cous, the margins weakly dentate, the adaxial surface asperous. Terminal stipules caducous leaving a conspicuous circular scar. Latex white. Infructes-cence axillary or borne from the stems. Fruit a multi-ple, to 5 cm diameter, pubescent, yellow when ma-ture, subtended by persistent calyx; the mesocarp yellow, edible. Seeds various per multiple fruit. Pu-bescence golden, very dense throughout the plant. Historically *Castillea ulei* has been exploited as a source of rubber latex. Distribution: Mexico to Peru, Bolivia, and Brazil.

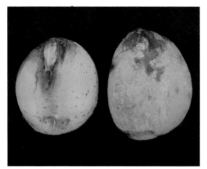

Moraceae - Castilla (2.9×; 12.96 × 10.04)

Clarisia R. & P. Trees to 40 m tall. Trunk cylindrical, lenticellate, reddish; some species with roots ex-posed at ground surface; strongly orange-reddish, es-pecially when rubbed or scraped. Terminal stipules reduced and caducous, not leaving a conspicuous circular scar. Latex white, abundant. Infructescence axillary. Fruit a drupe, to 3 cm long, red or yellowish when mature, the mesocarp yellow or green. Seeds one per fruit. Some species used for lumber. Distribu-tion: Mexico to Bolivia and Peru.

Moraceae - Clarisia (2.07×; 20.05 × 13.27)

Dorstenia L. Shrubs to 1.5 m tall. Leaves very thin or nearly transparent when dry, the base cordate. In-fructescence axillary, technically a swollen recepta-cle, to 4.8 cm diameter, flattened above or cup-shaped, asperous, green when mature, forming a multiple fruit. Seeds various per multiple fruit. Plants with short pubescence. Distribution: Mexico to Peru, but more speciose in the Old World.

Moraceae - Dorstenia (5.87×; 3.15 × 2.85)

Ficus L. Primarily hemiepiphytic or stranglers, but also trees up to 40 m tall. Leaves variable in size and shape but always with entire margins. Terminal stipule conspicuous, caducous, leaving a circular scar when falling. Latex generally white, abundant, sometimes without color or nearly absent. Infructescence axillary or borne from the stems, sessile or pedunculate, paired or solitary, forming the multiple fruit, known as a syconium, that is often subtended by a bract that is small or to a size sometimes covering the entire syco-nium. Fruit technically small achenes enclosed in the syconium, from 0.3 cm to 4.5 cm diameter, red, yellow, pink, purple, green, or a mix of those colors, or rarely black when mature. Some species with large promi-nent opening, or ostiole, at the apex of the synconium. Seeds tiny, numerous per syconium. Plants glabrous or pubescent. Tree species often with large flat and flying buttress roots; a few with stilt roots. Many spe-cies of this genus are considered "keystone species" because their fruits are abundant and available throughout the year, thus serving as food resources for many animal species. One species cultivated for edible fruit. Distribution: Southern United States, Mex-ico, and the Antilles to Paraguay and northern Argen-tina, but more speciose in the Old World, to 2300 m elevation.

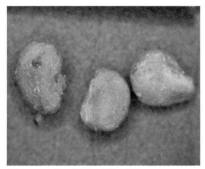

Moraceae - *Ficus* (12.49x; 1.92 × 1.32)

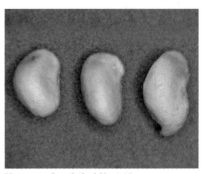

Moraceae - *Ficus* (9.63x; 2.92 × 1.64)

Helicostylis Tréc. Trees to 25 m tall. Leaves usually with serrate margins. Terminal stipule small and the scar not completely circular. Latex tan to yellowish. Infructescence axillary, pedunculate, or sessile, developing into a multiple fruit to 6 cm diameter, yellow and very smooth when mature. Seeds 1–14 embedded in the juicy yellow mesocarp of each multiple fruit. Pubescence golden-yellow to brown-rufescent, dense, short, sometimes scabrous. Distribution: Guianas, Brazil, Colombia, Bolivia, and Peru, to 2300 m elevation.

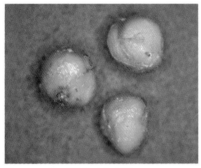

Moraceae - *Ficus* (15.13x; 1.4 × 1.32)

Moraceae - *Helicostylis* (2.52x; 13.11 × 8.57)

Maquira Aubl. Trees to 30 m tall. Leaves coriaceous and very glossy, especially on the adaxial surface. Terminal stipule small, caducous, leaving a circular scar when falling. Latex tan, rarely white. Infructescence axillary, sessile, the subtending bracts

Moraceae - *Ficus* (11.11x; 2.45 × 1.55)

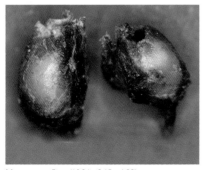

Moraceae - *Ficus* (18.94x; 2.15 × 1.29)

Moraceae - *Maquira* (1.78x; 24.35 × 14.52)

succulent; the calyx and stigmas persistent. Fruit a multiple, to 6 cm diameter, yellow or orange when mature. Seeds 1–5 per multiple fruit, embedded in a yellow or orange, juicy, fibrous mesocarp. Plants usually glabrous. Distribution: Nicaragua and Guianas to Bolivia and Peru, to 1800 m elevation.

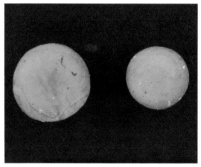

Moraceae - *Maquira* (2.48x; 10.9 × 10.6)

Naucleopsis Miq. Trees to 20 m tall, or sometimes shrubs. Leaves coriaceous, the adaxial surface glossy in some species. Terminal stipule conspicuous, persistent, late caducous, leaving a circular scar when falling. Latex yellow, somewhat watery. Infructescence axillary, sessile, forming a multiple fruit; the subtending bracts persistent, spiny, elongate, to 15 cm diameter, yellow when mature. Seeds 5–12 per multiple fruit, embedded in the yellow, fibrous mesocarp. Plants usually glabrous. Distribution: Guianas, Brazil, Peru, and Bolivia.

Moraceae - *Naucleopsis* (2.73x; 16.12 × 10.5)

Perebea Aubl. Trees to 15 m tall, or sometimes shrubs. Leaves with entire or dentate margins, the adaxial surface asperous or the base oblique in some species. Terminal stipule conspicuous, caducous, leaving a circular scar when falling. Latex tan to yellow. Infructescence axillary, forming a multiple, with 1–4 fruits grouped freely or tightly on a flat receptacle, to 5.5 cm diameter, yellow or red when mature, all subtended by bracts. Seeds one per fruit, embedded in the red mesocarp. Plants pubescent or glabrous. Distribution: Costa Rica to Bolivia and Peru.

Moraceae - *Perebea* (4.61x; 6.93 × 5.28)

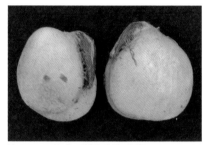

Moraceae - *Perebea* (3.85x; 8.49 × 8.02)

Pseudolmedia Tréc. Trees to 40 m tall. Terminal stipule conspicuous, caducous, leaving a circular scar when falling. Latex tan. Infructescence axillary, the calyx and stigma persistent. Fruit a drupe, to 3.5 cm diameter, red when mature, the mesocarp juicy, fleshy, red, sweet, and edible. Seeds one per fruit. Plants pubescent or glabrous. Distribution: Southern Mexico and Guianas, Venezuela, to Peru and Bolivia, to 2300 m elevation.

Moraceae - *Pseudolmedia* (5.31x; 8.15 × 5.34)

Sorocea A. St.-Hil. Shrubs or trees to 25 m tall. Leaves sometimes asperous, the margins weakly serrate. Terminal stipule caducous, leaving a scar that is not circular. Latex tan or white. Infructescence axillary and borne from the stems, the pedicel often swollen. Fruit a drupe, sessile, to 2 cm long, black when mature, subtended by persistent, succulent, gray, or red perianth; the mesocarp green, fleshy, carnose. Seeds one per fruit. Plants pubescent or glabrous. Distribution: Central America to Brazil, Peru, and Bolivia.

Moraceae - *Sorocea* (2.77x; 12.21 × 8.74)

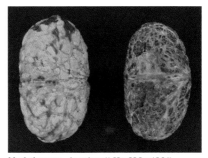

Myristicaceae - *Iryanthera* (1.82x; 23.2 × 13.26)

Trymatococcus P. & E. Trees to 18 m tall. Leaves relatively small, pubescent, asperous. Latex transparent, aqueous. Infructescence globose, axillary, developing into a multiple fruit, to 3.5 cm diameter, asperous, green when mature, crowned by the remains of staminate flowers. Seeds one per fruit. Distribution: Guianas and Venezuela to Brazil and Peru.

Osteophloeum Warb. Trees to 40 m tall. Resin from trunk transparent, acolorous, which differs from other members of the family. Leaves lightly coriaceous, the apex rounded. Infructescence axillary. Fruit to 3 cm diameter, conspicuously wider than long, green-yellow when mature. Seeds one per fruit, completely covered by red aril. Plants glabrescent. Some species harvested for their wood. Distribution: Panama to northern South America.

Moraceae - *Trymatococcus* (2.67x; 18.45 × 17.54)

Myristicaceae - *Osteophloeum* (1.81x; 22.46 × 16.14)

MYRISTICACEAE

Trees. Leaves simple, alternate, always entire. Trunk with red resin, sometimes transparent, aqueous in young individuals. Infructescence axillary or borne from the stems or trunk, rarely terminal. Fruit referred to as a capsule or a dehiscent berry, rarely indehiscent. Seed one per fruit, usually arillate. Some species with aerial roots, others with small buttresses. Plants glabrous or pubescent, the hairs simple, T-shaped, dendritic, or stellate. Branching conspicuously verticillate; referred to as "myristicaceous branching."

Iryanthera Warb. Trees to 30 m tall. Leaves weakly coriaceous, the tertiary venation inconspicuous, the midvein prominent. Infructescence borne from the stems or trunk. Fruit to 4.5 cm diameter, conspicuously wider than long, green-yellow when mature. Seeds one per fruit, with red aril, positioned transversely in fruit. Plants glabrous to glabrescent, the hairs T-shaped. Some species used for their wood. Distribution: Central America to Peru and Bolivia.

Otoba (DC.) Karst. Trees to 30 m tall. Resin from trunk red. Leaves with glaucous to lightly tomentose abaxial surface, the tertiary venation visible or inconspicuous. Infructescence axillary. Fruit to 4.5 cm diameter, usually globose, green when mature. Seeds one per fruit, covered with a white to tan aril. Plants pubescent, the hairs T-shaped. Distribution: Tropical America.

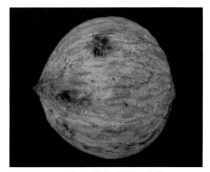

Myristicaceae - *Otoba* (2.43x; 20.17 × 19.43)

Virola Aubl. Trees to 35 m tall. Leaves usually with conspicuous secondary and tertiary venation, or sometimes inconspicuous; the lamina base strongly cordate in some species. Infructescence axillary or terminal. Fruit to 3.6 cm diameter, generally longer than wide, green to brown when mature. Seeds one per fruit, positioned longitudinally in the fruit, covered by a yellow, orange, or red aril. Plants pubescent, the hairs simple or stellate, short or long, dense or thin; rarely glabrous. Terminal stem and apical bud typically tomentose. Some species with edible seeds or harvested for their wood. Distribution: Central America to Peru and Bolivia, to 1500 m elevation.

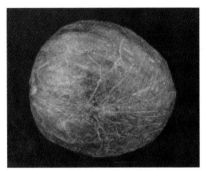

Myristicaceae - *Virola* (3.32x; 15.29 × 13.79)

Myristicaceae - *Virola* (1.96x; 16.49 × 14.32)

MYRSINACEAE

Shrubs or small trees. Leaves simple, alternate, with glandular punctations or with a covering of scales; the margins entire. Infructescence axillary, terminal, or borne from the stems or trunk. Fruit a drupe subtended by persistent, striate calyx; the mesocarp fleshy, juicy. Seeds one per fruit. Plants glabrous or pubescent.

Ardisia Sw. Shrubs or trees to 5 m tall. Leaves sometimes succulent, coriaceous, the secondary venation weakly decurrent; the margins undulate-crenate; the abaxial laminar surface with conspicuous small, brown punctations. Infructescence terminal, axillary, or borne from the trunk. Fruit to 1.5 cm diameter,

black when mature, the mesocarp green to dark purple. Plants glabrous to weakly pubescent. Distribution: Central America to Peru and Bolivia, but more speciose in Asia.

Myrsinaceae - *Ardisia* (6.4x; 4.31 × 4.23)

Cybianthus C. Martius. Shrubs or trees to 8 m tall. Leaves with brown and black punctations on the abaxial surface. Infructescence axillary or borne from the trunk. Fruit to 1.4 cm diameter, black when mature, subtended by persistent calyx. Plants glabrous, tomentose, or with scales. Distribution: Central America to Peru and Bolivia, to 2500 m elevation.

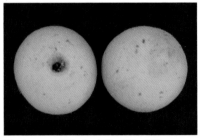

Myrsinaceae - *Cybianthus* (4.25x; 6.97 × 6.79)

Stylogyne A. DC. Shrubs or trees to 10 m tall. Leaves with very dense and conspicuous glandular punctations; the secondary and tertiary venation indistinguishable and inconspicuous; the midvein prominent below. Infructescence axillary, terminal, or borne from the trunk, the peduncle and pedicels reddish. Fruit to 1.5 cm diameter, black when mature, subtended by persistent calyx. Distribution: Mexico to Peru and Bolivia.

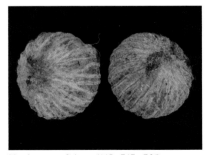

Myrsinaceae - *Stylogyne* (4.15x; 7.17 × 7.04)

MYRTACEAE

Shrubs and trees, usually aromatic. Trunks often smooth or with papyraceous, defoliating bark, usually whitish or reddish in color. Leaves simple, opposite, the margins entire; the secondary veins often anastomosing at the margins; the lamina with conspicuous translucent punctations. Infructescence axillary, terminal, or borne from the stems or trunk. Fruit a drupe or berry, crowned by persistent calyx.

Calyptranthes Sw. Trees to 15 m tall. Leaves with undulate margins; the secondary venation very fine and decurrent, nearly perpendicular to the midvein; the lamina with conspicuous or inconspicuous translucent punctations. Infructescence axillary, terminal, or borne from the trunk. Fruit a berry, to 6 cm diameter, crowned by a persistent calyx and sometimes marked with prominent longitudinal lines and grooves, yellow, dark purple to black when mature; the mesocarp white or transparent; the exocarp with translucent punctations. Seeds 1–2 per fruit. Plants pubescent, the hairs 2-branched. Distribution: Southern United States to Peru and Bolivia.

Myrtaceae - *Calyptranthes* (2.52x; 10.92 × 8.15)

Myrtaceae - *Calyptranthes* (5.19x; 5.8 × 5.37)

Campomanesia R. & P. Shrubs or trees to 10 m tall. Leaves strongly aromatic when crushed, the secondary venation reticulate, anastomosing only in the upper half of the lamina, usually inconspicuously. Punctations reddish-transparent and conspicuous at the apices of the branches and on the abaxial laminar surface, but hidden on the adaxial surface. Infructescence axillary. Fruit a berry, to 3.3 cm long, crowned by persistent calyx, yellow when mature, with very dense reddish-transparent punctations on the pedicel and exocarp; the mesocarp

transparent or white. Seeds 2–7 per fruit. Distribution: Tropical South America.

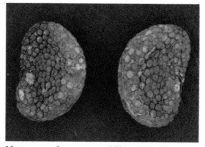

Myrtaceae - *Campomanesia* (2.73x; 12.79 × 8.91)

Eugenia L. Shrubs or trees to 25 m tall. Bark papyraceous on trunks and stems in some species. Leaves often with anastomosing secondary venation, the veins often conspicuously decurrent in the upper half, the translucent punctations very conspicuous or inconspicuous at a simple glance, with small black punctations on the abaxial laminar surface. Infructescence axillary, terminal, or borne from the stems. Fruit a berry, pedunculate or sessile, to 5.5 cm diameter, red, dark purple, yellow, or orange when mature, terminated by a persistent stigma and crowned by persistent calyx; the exocarp with glandular punctations. Seeds 1–4 per fruit. Plants pubescent or glabrescent. Some species with edible fruit. Distribution: Tropical and subtropical America and Asia to 1400 m elevation.

Myrtaceae - *Eugenia* (1.77x; 27.13 × 18.56)

Myrtaceae - *Eugenia* (3.13x; 15.3 × 7.2)

Myrtaceae - *Eugenia* (2.16x; 14.17 × 12.25)

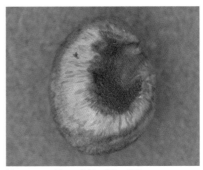

Myrtaceae - *Myrcia* (5.12x; 9.11 × 7.19)

Myrcia DC. Ex. Guillemin. Shrubs or trees to 15 m tall. Leaves with anastomosing or very fine secondary venation, and usually with a conspicuous marginal vein; the translucent punctations conspicuous or inconspicuous at a simple glance. Infructescence axillary or borne from the trunk. Fruit a berry to 3 cm diameter, dark purple or black when mature, terminated by a persistent stigma and crowned by a persistent calyx; the exocarp with glandular punctations; the mesocarp transparent white to dark purple. Seeds 1–3 per fruit. Plants with dense pubescence of T-shaped hairs. Distribution: Tropical America, to 2000 m elevation.

Myrciaria O. Berg. Trees to 15 m tall. Leaves with very fine secondary venation; the translucent punctations inconspicuous at a simple glance. Infructescence axillary or borne from the stems. Fruit a sessile berry to 1 cm diameter, red when mature, crowned by a persistent calyx. Seeds 1–2 per fruit. Some species with edible fruits. Distribution: Mexico to Argentina, to 1400 m elevation.

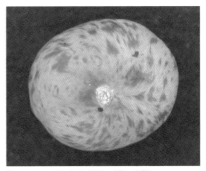

Myrtaceae - *Myrciaria* (7.62x; 6.74 × 5.72)

Myrtaceae - *Myrcia* (7.2x; 5.87 × 3.76)

Psidium L. Trees to 8 m tall. Some species with square branch apices formed in part by small papyraceous wings. Leaves with prominent secondary venation on abaxial surface, the veins clearly anastomosing; the translucent punctations rather inconspicuous. Infructescence axillary. Fruit a berry, to 5 cm

Myrtaceae - *Myrcia* (3.13x; 19.09 × 8.78)

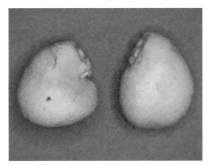

Myrtaceae - *Psidium* (5.61x; 5.81 × 4.92)

diameter, yellow or red when mature, terminated by a persistent stigma and crowned by a persistent calyx. Seeds various per fruit. Plants pubescent. Some species with edible fruits. Distribution: Mexico to Paraguay, to 2500 m elevation.

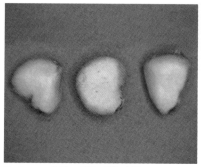

Myrtaceae - *Psidium* (5.71×; 3.35 × 3.06)

NYCTAGINACEAE

Shrubs and trees, rarely lianas. Terminal shoot buds typically ferrugineous, with short reddish-brown hairs. Leaves simple, entire, alternate, but often confusingly appearing subopposite to opposite; drying black in most species. Infructescence axillary, terminal, or borne from the trunk. Fruit often referred to as a drupe, but technically an anthocarp due to the fusion of perianth and ovary wall. Seeds one per fruit. Plants glabrous, glabrescent, or sometimes tomentose pubescent, and sometimes with spines.

Neea R. & P. Shrubs, or trees to 25 m tall. Terminal shoot buds usually ferrugineous. Leaves weakly succulent, often anisophyllous, the venation somewhat inconspicuous. Infructescence often with red to dark purple peduncle. Fruits pedicellate or sessile, to 1.8 cm long, pink, dark purple to black when mature, terminated by persistent stigma and crowned by persistent calyx; the mesocarp transparent to dark purple. Plants mostly glabrous, rarely pubescent. Distribution: Costa Rica to Argentina.

Nyctaginaceae - *Neea* (2.65×; 13.96 × 7.88)

Nyctaginaceae - *Neea* (5.44×; 9.3 × 2.94)

OCHNACEAE

Shrubs or trees, with one genus of herbs. Leaves usually simple and alternate, but compound in one genus; the margins usually strongly serrate. Stipules present, very conspicuous. Infructescence axillary or terminal. Fruit a drupe or a capsule. Seeds winged or not. Plants usually glabrous.

Ouratea Aubl. Shrubs, or trees to 20 m tall. Leaves coriaceous, the secondary venation very fine and arching from midvein to apex, irregular, not arriving to the margin; the margins finely dentate, the teeth very strong, sometimes like spines; the adaxial lami-

Nyctaginaceae - *Neea* (2.67×; 14.6 × 7.42)

Ochnaceae - *Ouratea* (3.3×; 11.79 × 7.66)

nar surface glossy. Stipules caducous, leaving a circular scar at each node. Infructescence axillary or terminal, the peduncle red. Fruit a glandarium, consisting of 1–6 drupes attached to the gynophore, a red carnose receptacle; each drupe 1.7 cm long, black when mature. Seeds one per drupe. Distribution: Mexico to Peru and Bolivia, also in the Old World.

OLACACEAE

Shrubs and trees, rarely lianas. Latex present from the leaf veins, aqueous white. Leaves simple, alternate, entire; the tertiary venation very fine and parallel. Infructescence axillary. Fruit a drupe. Seeds one per fruit.

Chaunochiton Benth. Trees to 30 m tall. Leaves weakly coriaceous. Infructescence axillary or terminal. Fruit to 1.7 cm diameter, brown when mature, situated at center of large persistent calyx that forms a round wing, to 13 cm diameter, green when immature, brown when mature. Distribution: Costa Rica to Peru and Bolivia.

Olacaeae - *Chaunochiton* (0.47×; 103.05 × 92.2)

Heisteria Jacq. Shrubs or trees to 25 m tall, or lianas. Leaves weakly coriaceous, the adaxial laminar surface glossy, the petioles elongate, twisted, reflexed, swollen on the upper half. Infructescence usually axillary, sometimes borne from the trunk. Fruit to

1.4 cm long, white or red when mature; subtended by persistent calyx that forms a small round wing to 5 cm diameter, or that surrounds the fruit, red or green when mature. Distribution: Mexico to Peru and Bolivia, also in Africa.

Minquartia Aubl. Trees to 35 m tall. Trunk sometimes fenestrate, especially in mature individuals. Latex aqueous white in the leaves when young, especially from midvein and secondary veins. Leaves with very fine tertiary venation, perpendicular to the secondary venation that is also finely parallel; the petiole elongate with apical pulvinus. Fruit to 2.6 cm long, black when mature, with white latex, especially when immature. Plants pubescent, the hairs stellate. Valued for extremely strong, dense wood that is resistant to moisture and often used to make tool handles. Distribution: Central America to Peru and Bolivia.

Olacaeae - *Minquartia* (2.19×; 22.32 × 12.71)

ONAGRACEAE

Herbs and shrubs. Leaves simple, alternate, opposite or verticillate. Stipules, if present, early caducous. Fruit a capsule or berry.

Fuchsia L. Lianas, shrubs, or small trees, sometimes hemiepiphytic. Branches with reddish exfoliating bark. Leaves opposite or verticillate. Stipules inconspicuous and caducous. Infructescence axillary or terminal. Fruit a berry to 2 cm, usually round, red to

Olacaeae - *Heisteria* (4.19×; 10.58 × 6.64)

Onagraceae - *Fuschia* (34.6×; 0.95 × 0.47)

dark purple when mature. Seeds small, many per fruit. Some species with edible fruits. Distribution: Mexico to Chile, the West Indies and Tahiti, with one species in New Zealand; to 3200 m elevation.

Ludwigia L. Herbs or shrubs to 2 m tall, sometimes semiaquatic. Leaves usually alternate. Infructescence axillary. Stigma persistent. Fruit a capsule, to 3.5 cm long or 1 cm diameter, brown when mature. Seeds various per fruit. Some species, especially the herbs, with hollow, square, or winged stems, or papyraceous bark. Plants usually pubescent or glabrescent. Growing in open humid sites, sometimes in wetlands. Distribution: North America to Argentina, also in the Old World.

Onagraceae - *Ludwigia* (32.17x; 0.86 × 0.33)

Onagraceae - *Ludwigia* (17.04x; 1.24 × 1.13)

OPILIACEAE

Trees, at least in our area, also shrubs and lianas. Leaves simple, alternate, the margins entire. Infructescence axillary or borne from the trunk. Fruit a drupe. Seed one per fruit. Plants glabrous.

Agonandra Miers ex Hook. f. Trees to 30 m tall. Latex cream-colored. Leaves coriaceous, the secondary venation finely parallel and not easily differentiated from the tertiary venation. Fruit a drupe, to 2.9 cm long, yellow or orange when mature; the mesocarp fibrous and sticky. Seeds one per fruit. Distribution: Mexico to Argentina.

Opiliaceae - *Agonandra* (1.47x; 18.2 × 13.88)

ORCHIDACEAE

Herbs, usually epiphytic, sometimes hemiepiphytic, terrestrial, or aquatic. Probably the most diverse family of flowering plants in the world. Characterized by the presence of pseudobulbs in the majority of the genera. Roots aerial, usually white. Leaves often succulent. Fruit a 3-valved capsule. Seeds microscopic, numerous per fruit, often winged. The Orchidaceae is a diverse family of herbs that are poorly represented in this field guide where we present a small set of images to illustrate some of the seed morphology present in the family. Although most orchid seeds are often winged, tan to brown when mature, and microscopic, seeds of the genus *Vanilla* are round, black, and more easily visible to the human eye. We present five representative genera found in the Amazon, *Elleanthus*, *Erythrodes*, *Habenaria*, *Huntleya*, and *Vanilla* .

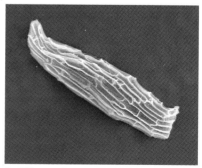

Orchidaceae - *Elleanthus* (137.57x; 0.45 × 0.11)

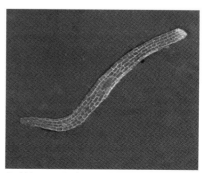

Orchidaceae - *Erythrodes* (57.14x; 1.09 × 0.11)

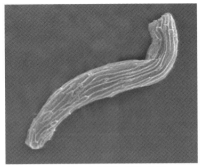

Orchidaceae - *Habenaria* (93.12x; 0.72 × 0.11)

Orchidaceae - *Huntleya* (198.94x; 0.3 × 0.09)

Orchidaceae - *Vanilla* (115.34x; 0.39 × 0.31)

OXALIDACEAE

Herbs and shrubs, sometimes trees. Leaves alternate, compound, trifoliate, paripinnate, or imparipinnate. Infructescence axillary. Fruit a berry or capsule.

Averrhoa L. Shrubs, or small trees to 4 m tall. Leaves compound, imparipinnate, the leaflets opposite or subopposite, generally smaller toward the base. Infructescence borne from the stems. Fruit a berry with 4–6 lobes, to 17 cm long, yellow when mature; the mesocarp juicy, abundant, edible. Seeds various per fruit. Cultivated throughout the tropics for edible fruit (star fruit or carambola). Distribution: Pantropical.

Oxalidaceae - *Averrhoa* (3.26x; 10.45 × 4.87)

PASSIFLORACEAE

Lianas, rarely shrubs. Tendrils simple and axillary in lianas, or absent in shrubs. Leaves simple, alternate, rarely compound, the margins entire or lobed, the petioles and laminar surfaces often with glands. Infructescence axillary. Fruit a berry. Seeds various per fruit.

Dilkea Mast. Shrubs or lianas. Leaves elliptic, weakly clustered at the extremes of the main stem or branches; the margins entire; the secondary veins pinnate; the apex acuminate. Infructescence axillary or terminal. Fruit a berry to 5 cm diameter, yellow when mature, apiculate at the apex, the apicule often

Passifloraceae - *Dilkea* (2.54x; 15.42 × 8.29)

sharp to the touch; the mesocarp white. Seeds various per fruit but fewer than in *Passiflora*. Distribution: Colombia to Peru and Bolivia.

Passiflora L. Lianas or climbing herbs. Tendrils simple, axillary. Leaves simple or compound, entire or lobed; the margins serrate or entire; the petiole base and base of the adaxial laminar surface often with one or two pairs of glands. Stipules present in pairs at each node. Infructescence axillary. Fruit to 9.3 cm diameter, green, yellow, red, orange, or rarely black when mature, solitary or in pairs; subtended by persistent corolla and glandular calyx; the exocarp to 1 cm thick; the mesocarp white, mealy. Seeds numerous, surrounded by a sweet, transparent, sticky aril. Plants pubescent or glabrous. Widely cultivated for edible fruits. Distribution: Southern United States, Mexico, Central America to Paraguay, also in the Old World.

Passifloraceae - *Passiflora* (7.3×; 5.99 × 3.68)

Passifloraceae - *Passiflora* (3.15×; 7.72 × 5.64)

PHYTOLACCACEAE

Herbs, shrubs, trees, or rarely lianas. Leaves simple, alternate, the margins entire. Stipules very small or absent. Infructescence axillary or terminal. Fruit a samara, drupe, or berry. Plants glabrous or pubescent. Sometimes spiny.

Gallesia Casar. Trees to 40 m tall, with a strong garlic odor. Trunk irregular, yellowish-cream in color. Leaves with weak yellow venation. Infructescence

terminal. Fruit a samara, to 3.5 cm long, subtended by persistent calyx. Distribution: Brazil, Peru, and Bolivia, to 1300 m elevation.

Phytolaccaceae - *Gallesia* (1.8×; 28.47 × 12.24)

Petiveria L. Herbs less than 1 m tall, sometimes aromatic. Leaves alternate, the margins entire. Infructescence terminal or axillary to 20 cm long. Fruit a two-lobed achene to 1 cm long, each lobe bearing hooks or uncinate bristles at the apex; brown when mature. Distribution: Southern United States to Argentina.

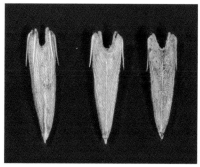

Phytolaccaceae - *Petiveria* (3.58×; 10.8 × 3.02)

Phytolacca L. Shrubs to 2 m tall, rarely trees. Stems yellow, rather succulent. Leaves very thin and fragile when dry. Infructescence terminal, the pedicels pink to red. Fruit a berry, to 0.8 cm diameter, black when mature. Seeds 5–16 per fruit. Distribution: Mexico to Peru and Bolivia, also in Asia and Africa, to 1800 m elevation.

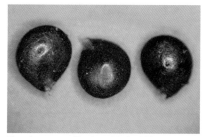

Phytolaccaceae - *Phytolacca* (10.48×; 2.22 × 1.87)

Seguieria Loefl. Lianas. Stems with curved or erect spines. Leaves usually with elongate petioles. Infructescence terminal. Fruit a samara, to 3.5 cm long, subtended by persistent calyx, very similar to *Gallesia*. Distribution: Tropical and subtropical America.

Phytolaccaceae - *Seguieria* (1.4x; 32.12 × 13.56)

Trichostigma A. Rich. Lianas. Infructescence axillary. Fruit a berry, to 0.9 cm diameter, black when mature. Seeds few per fruit. Distribution: Tropical America.

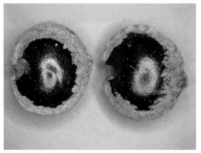

Phytolaccaceae - *Trichostigma* (6.08x; 5.34 × 4.78)

PIPERACEAE

Herbs and shrubs, erect or lianescent, epiphytes or rarely taking on the form of small trees, aromatic when rubbed or crushed. Nodes conspicuous, swollen along the stem. Leaves simple, opposite or alternate, the margins entire. Stipules present or absent. Infructescence axillary, opposite the leaf, erect or pendulous. Fruit a drupe, many per infructescence. Seeds one per fruit.

Piper L. Shrubs to 5 m tall. Leaves generally with asymmetric base, sometimes peltate; the petiole base sheathing the stem; the laminar surface asperous in some species. Infructescence spicate, axillary to 30 cm long. Fruit a drupe, sessile, to 0.5 cm long, green or yellowish, generally aromatic to malodorous when mature; terminated by persistent stigma. Seed one per fruit. Plants pubescent or glabrous. One species

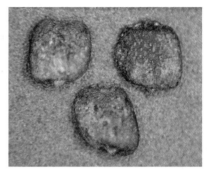

Piperaceae - *Piper* (14.18x; 1.77 × 1.64)

Piperaceae - *Piper* (11.11x; 2.03 × 1.58)

Piperaceae - *Piper* (10.48x; 3.47 × 1.86)

Piperaceae - *Piper* (9.74x; 2.61 × 1.59)

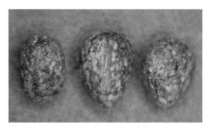

Piperaceae - *Piper* (11.96x; 2.13 × 1.54)

widely cultivated for its seed, which is used as the source of white and black pepper. Many species used in folk medicine. Distribution: Tropical America, Asia, and Africa, to 3000 m elevation.

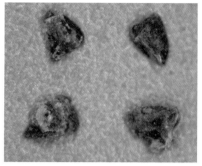

Piperaceae - Piper (31.64x; 0.6 × 0.49)

POACEAE

Herbs, erect, prostrate, or lianescent, rarely taking on the form of small trees; often rhizomatous or stoloniferous. Stems, referred to as culms, usually with hollow internodes and solid, swollen nodes. Leaves simple, alternate or basal, linear, the petiole base sheathing the stem; the margins entire or sometimes finely serrate or minutely spinate; the venation fine and parallel. Infructescence axillary or terminal, in the form of one or more spikelets. Fruit a caryopsis. Seeds one per fruit. Known as the grass family, the Poaceae is widespread, diverse, and separated into six subfamilies. Only a selection of common genera are presented here.

Acroceras Stapf. Herbs to 1 m tall. Leaf margins sometimes with small spines, the venation fine and parallel. Infructescence terminal. Fruit carnose, to 0.6 cm long, sometimes white and glossy when mature. Plants with soft, dense pubescence throughout. Distribution: Mexico to Paraguay, also in the Old World.

Poaceae - Acroceras (7.78x; 5.73 × 1.88)

Axonopus P. Beauv. Herbs to 0.8 m tall. Leaf margin weakly spinose. Infructescence terminal. Fruit to 0.3 cm diameter, whitish when mature. Plants glabrescent; growing in large groups. Distribution: Tropics of the New and Old World.

Poaceae - Axonopus (19.26x; 1.95 × 0.71)

Guadua (Kunth) Hackel. Semi-woody herbs to 18 m tall. Rhizomes large and aggressive. Stems with pair of spines at each node, curving downward, or erect. Leaves with a pseudopetiole, the margins with up to three rows of small swollen spines. Infructescence axillary or terminal. Fruit to 3 cm long, reddish when mature. Plants with short, dense pubescence. Plants generally covering vast areas of past disturbance. Distribution: Mexico to Argentina.

Poaceae - Guadua (2.67x; 17.98 × 5.4)

Lasiacis (Griseb.) Hitchc. Sub-woody herbs to 2 m tall, erect or lianescent. Leaf margin sometimes with small spines. Fruit to 0.3 cm diameter, black and glossy when mature. Distribution: Mexico to Argentina.

Poaceae - Lasiacis (7.78x; 3.66 × 2.61)

Olyra L. Subwoody herbs to 2 m tall, erect or lianescent. Leaf margins sometimes with small spines. Infructescence terminal. Fruit to 1 cm long, subtended by persistent glumes. Plants with short pubescence that is sometimes white. Distribution: Southern United States to Argentina, one species in Africa.

Poaceae - *Olyra* (4.3x; 9.41 × 3.99)

Poaceae - *Olyra* (5.31x; 5.88 × 3.23)

Panicum L. Herbs to 2 m tall, erect or prostrate. Stems smooth, finely striate. Leaf margins ciliate or sometimes with small spines. Infructescence terminal. Fruit to 0.3 cm long, whitish when mature, sometimes glossy. Plants pubescent or glabrous. Distribution: United States, Mexico to Paraguay, also in Africa.

Poaceae - *Panicum* (14.29x; 1.79 × 1.26)

Paspalum L. Herbs to 2 m tall, erect. Stems smooth, finely striate. Leaves linear, weakly pubescent, the margins sometimes weakly spinose. Infructescence terminal. Fruit to 0.4 cm long, sometimes weakly striate or glossy. Plants pubescent or glabrous. Distribution: Southern United States and Mexico to Paraguay and Argentina, also in Africa.

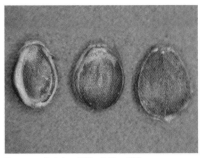

Poaceae - *Paspalum* (9.95x; 2.47 × 1.99)

Poaceae - *Paspalum* (16.93x; 1.5 × 1.01)

Pharus P. Browne. Herbs to 1 m tall. Leaves sometimes with ciliate or minutely spinose margins; the secondary venations fine and parallel; midvein absent. Infructescence terminal. Fruit asperous and adhesive, to 2.3 cm long, brown when mature, subtended and often completely covered by long, persistent glumes. Plants with short pubescence throughout. Distribution: Southern United States, Mexico, Central America to Peru.

Poaceae - *Pharus* (3.64x; 13.31 × 1.74)

POLYGALACEAE

Lianas, herbs, and shrubs, or rarely small trees. Leaves simple, alternate, the margins entire. Stipules present or absent. Infructescence axillary or terminal. Fruits variable; a capsule, berry, drupe, or samara.

Moutabea Aubl. Shrubs to 5 m, sometimes prostrate, or lianas. Stems with inconspicuous, minute spines. Leaves succulent, the secondary venation inconspicuous. Infructescence axillary. Fruit a berry, to 4 cm diameter, yellow or orange when mature, weakly pubescent; the exocarp coriaceous to somewhat indurate; the mesocarp cream-colored when mature. Seeds up to five per fruit. Distribution: Brazil, Peru, and Bolivia.

Polygalaceae - *Moutabea* (2.56x; 18.88 × 11.03)

Polygala L. Herbs to 1 m; rarely shrubs or small trees. Leaves thin, nearly translucent when dry, the secondary veins anastomosing at the margins. Infructescence terminal. Fruit a capsule, to 1.6 cm long, green when mature, subtended and often covered by persistent calyx. Seeds two per fruit, with a white aril at the base. Plants with short, dense pubescence. Distribution: North America to Peru and Bolivia, also in the Old World.

Polygalaceae - *Polygala* (6.62x; 9.14 × 3.99)

Securidaca L. Lianas or herbaceous vines; sometimes small trees. Stems sometimes compressed or canaliculate. Some species with raised glands on the stems or at petiole insertion, or with translucent to reddish sap. Infructescence axillary or terminal. Leaves thin or coriaceous, glabrous or pubescent, the secondary venation conspicuous. Fruit a samara with one nerved, apical wing, to 7 cm long, brown when mature; sometimes with a small secondary wing. Seeds one per fruit. Shoots and bark of some species used as medicine. Distribution: Pantropical, to 1000 m elevation.

Polygalaceae - *Securidaca* (0.97x; 62.95 × 19.78)

POLYGONACEAE

Shrubs, trees, and lianas. Leaves simple, alternate, the margins entire. The family characterized by a terminal or axillary sheathing stipule (ocrea) at each node, which leaves a circular scar upon falling. Fruit an achene covered by persistent perianth. Plants glabrous or pubescent.

Coccoloba P. Browne. Lianas, shrubs, or trees to 18 m. Leaves with anastomosing venation, the laminar base with glands in some species. Ocrea completely surrounding the stem. Infructescence axillary or terminal. Fruit to 1.8 cm diameter, red or black when mature. Seeds one per fruit. In some species the fruit opens to expose the white-arillate seed. Plants with dense, white pubescence. Distribution: Central America to Peru and Bolivia, to 2200 m elevation.

Polygonaceae - *Coccoloba* (6.24x; 6.36 × 4.36)

Polygonum L. Herbs or shrubs to 1 m, succulent and sometimes aquatic. Ocrea completely surrounding the stem. Infructescence axillary or terminal. Fruit to 0.5 cm long. Plants pubescent or glabrous. Distribution: North and South America, and in the Old World.

Polygonaceae - *Polygonum* (9.84×; 2.37 × 1.66)

Ruprechtia C.A. Mey. Shrubs or trees to 25 m tall. Infructescence terminal or axillary. Fruit a sessile achene to 0.5 cm diameter, with three wings to 2.5 cm long formed from the calyx lobes; pubescent or glabrescent, pink when immature, tan to brown when mature. Distribution: Mexico to Uruguay and Argentina, to 2200 m elevation.

Polygonaceae - *Ruprechtia* (1.33×; 17.78 × 3.02)

Polygonaceae - *Ruprechtia* (0.87×; 63.98 × 10.15)

Triplaris Loefl. ex. L. Trees to 20 m. Leaves with weakly anastomosing venation, the lamina with glandular punctations. Ocrea completely surrounding the stem. Infructescence axillary or terminal. Fruit a sessile achene to 1 cm diameter, with three wings to 4 cm long formed from the calyx lobes; pubescent or glabrous, red, showy when immature, brown when mature. Stems of some species are hollow and sometimes associated with aggressive ants. Plants often with golden-brown pubescence. Distribution: Southern Mexico to Peru and Bolivia.

Polygonaceae - *Triplaris* (1.88×; 31.85 × 5.11)

PROTEACEAE

Shrubs and trees. Leaves alternate, simple or compound, usually coriaceous. Infructescence axillary or terminal. Fruit a follicle, samara, or drupe. Seeds winged or not.

Roupala Aubl. Shrubs or trees to 25 m tall. Leaves strongly dimorphic, with simple and compound leaves on the same individual; the leaflets very asymmetric with entire or strongly serrate margins; the petioles elongate when the leaves are simple. Infructescence axillary or terminal. Fruit a follicle, to 5 cm long, brown when mature, the exocarp woody. Seeds winged. Genus easy to recognize in the field by the strongly nauseating odor of the bark. Distribution: Mexico to Argentina, to 2500 m elevation.

Proteacea - *Roupala* (1.35×; 38.59 × 17.19)

QUIINACEAE

Shrubs and trees. Leaves simple or compound, opposite or 3- to 4-verticillate, the margins entire or finely serrate. Stipules axillary, conspicuous, four per node. Infructescence axillary or terminal. Fruit a berry. Seeds 1–4 per fruit. Plants generally glabrous.

Quiina Aubl. Trees to 20 m. Leaves simple, coriaceous; the adaxial laminar surface glossy; the secondary venation fine and curved toward the apex; the tertiary venation inconspicuous. Stipules axillary, persistent, in one or two pairs, slender, sometimes very long and conspicuous. Infructescence axillary or terminal. Fruit a berry, longitudinally striate, to 5 cm long, yellow to cherry red or black when mature, subtended by persistent calyx; the mesocarp fibrous, acidic, malodorous. Plants generally glabrous or sometimes with short pubescence. Distribution: Central America to Peru and Bolivia, to 1600 m elevation.

Quiinaceae - *Quiina* (3.41×; 13.45 × 8.35)

RHAMNACEAE

Lianas, shrubs, or trees. Leaves simple, alternate or opposite, the margins entire or serrate, the secondary venation conspicuously parallel, pinnate, or 3-veined from the base. Stipules very small, caducous. Infructescence axillary or terminal. Fruit a drupe or capsule. Plants sometimes with spines or with glands on the leaves or tendrils.

Gouania Jacq. Lianas. Tendrils axillary or terminal, rolled from the end like a butterfly tongue. Leaves alternate, the margins serrate or entire, the laminar base with a pair of glands. Infructescence axillary or terminal. Fruit a winged capsule with three valves, to 1.6 cm diameter, sometimes dehiscing as a schizocarp. Plants pubescent or glabrous. Distribution: Southern United States (Florida) to Peru and Bolivia.

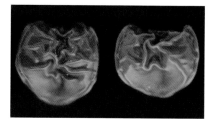

Rhamnaceae - *Gouania* (10.63×; 2.76 × 2.6)

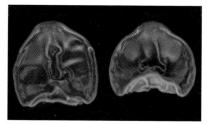

Rhamnaceae - *Gouania* (10.79×; 2.68 × 2.56)

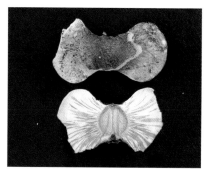

Rhamnaceae - *Gouania* (2.67×; 14.44 × 5)

Zizyphus Mill. Shrubs or trees to 30 m tall. Leaves simple and alternate, coriaceous, strongly 3-veined, the margins entire to serrate, the base weakly asymmetric. Infructescence axillary. Fruit a drupe to 2.5 cm long. Many species with spines on the stems, especially in dry forests, but the one species in the Amazon lacking spines. Distribution: Tropical America, but more speciose in the Old World.

Rhamnaceae - *Zizyphus* (3.07×; 19.66 × 15.38)

RHIZOPHORACEAE

Trees, rarely shrubs. Leaves simple, alternate, or opposite. Stipules present, usually caducous leaving a conspicuous scar. Infructescence axillary. Fruit a berry or a capsule. Family recognized by the genera that form mangroves, arising from pneumatophoric roots or aerial roots. It is sometimes confused with the Rubiaceae, especially in the Amazon.

Cassiporea Aubl. Shrubs or small trees. Leaves opposite. Stipules caducous leaving an intrapetiolar scar between the stem and petiole base. Infructescence axillary. Fruit a drupe to 1.5 cm long, yellow when mature, with persistent calyx and stigma. Seeds one per fruit. Distribution: Tropical America, but more speciose in Old World tropics.

Rhizophoraceae - *Cassiporea* (4.04x; 8.93 × 3.64)

ROSACEAE

Herbs, shrubs, or trees. Leaves simple or compound, pinnate or trifoliate, alternate or grouped at branch apices; the base usually cordate, the margins serrate. Stipules persistent or caducous. Infructescence axillary or terminal. Fruit variable; drupe, aggregate, pome, or berry.

Prunus L. Shrubs or trees to 25 m tall. Leaves simple, alternate, often with glands on the laminar surface. Stipules caducous. Infructescence axillary. Fruit a drupe. Seeds one per fruit. Many species cultivated for their edible fruits (i.e., almond, cherry, peach, and plum). Distribution: Tropical America, better repre-

Rosaceae - *Prunus* (2.35x; 12.21 × 10.41)

sented in the Northern Hemisphere and the Old World, to 3600 m elevation. One species in the Amazon; generally more common in montane forests.

RUBIACEAE

Herbs, shrubs, trees, and lianas, rarely epiphytes. Leaves simple, opposite, sometimes verticillate and entire. Stipules conspicuous, axillary or terminal. Infructescence axillary or terminal. Fruit a capsule, berry, or drupe. Seeds two, few, to many per fruit, sometimes with wings. Some genera with fenestrate trunks, papyraceous bark, or spines; a few species myrmecophilous.

Alibertia A. Rich. Ex DC. Shrubs or trees to 10 m tall. Leaves more or less small with stipules fused at base. Infructescence terminal. Fruit a sessile berry to 3.5 cm diameter, yellow or black when mature; pubescent or glabrous; crowned by conspicuous, persistent, tubular calyx; the mesocarp black. Seeds many per fruit. Plants pubescent or glabrous. Distribution: Tropical America.

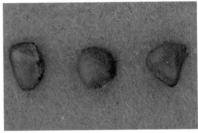

Rubiaceae - *Alibertia* (2.8x; 5.43 × 4.91)

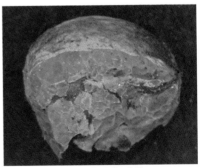

Rubiaceae - *Alibertia* (6.65x; 8.12 × 6.64)

Amaioua Aubl. Shrubs or trees to 10 m tall. Leaves weakly clustered at apices of the branches, the tertiary venation inconspicuous. Infructescence axillary. Fruit a berry, to 1.7 cm long, dark purple to blackish when mature. Seeds many per fruit. In some species the trunk fenestrate, in others the branches with defoliating bark. Plants with dense, short pubescence. Distribution: Mexico to Peru and Bolivia, to 1600 m elevation.

Rubiaceae - *Amaioua* (8.52×; 3.9 × 3.5)

Bathysa C. Presl. Trees to 8 m tall. Leaves coriaceous, sometimes 3-verticillate; the secondary venation anastomosing; the tertiary venation perpendicular to the secondary. Infructescence terminal. Fruit a capsule, to 2 cm long, brown when mature. Plants with dense, short pubescence. Distribution: Venezuela to Peru and Bolivia, to 1700 m elevation.

Bertiera Aubl. Shrubs to 5 m tall. Infructescence terminal. Fruit a berry, sessile or weakly pedicillate, to 1 cm diameter, light blue when mature, blue or nearly black when mature. Seeds 2–18 per fruit, organized in groups of 2–3. Plants with dense, short pubescence. Distribution: Tropical America, but more speciose in Africa.

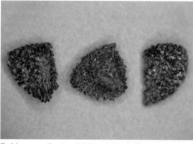

Rubiaceae - *Bertiera* (11.01×; 1.86 × 1.69)

Borojoa Cuatrec. Trees to 6 m tall. Infructescence terminal. Fruit a sessile berry, to 6.5 cm diameter; the exocarp to 0.5 cm thick; the mesocarp fleshy. Seeds many per fruit. Plants with dense, short pubescence. Distribution: Panama to Peru and Bolivia.

Rubiaceae - *Borojoa* (4.25×; 6.72 × 5.27)

Botryarrhena Ducke. Trees to 30 m tall. Leaves relatively large. Stipules surrounding the branch. Infructescence terminal. Fruit a berry, more or less bilocular, to 2 cm diameter; the exocarp woody. Distribution: Venezuela to the northeast of Peru and Brazil.

Rubiaceae - *Botryarrhena* (2.65×; 11.36 × 7.68)

Calycophyllum DC. Trees to 35 m tall. Bark papyraceous, reddish, defoliating, leaving the trunk very smooth and reddish to greenish. Leaves clustered at the branch apices. Stipules generally caducous. Infructescence terminal, with persistent bracts at the base; one calyx lobe highly developed, elongate, and persistent in some species. Fruit a bivalved capsule, to 1.3 cm long, brown when mature. Seeds winged, tiny, many per fruit. Plants with dense, short, golden pubescence, or glabrescent. Most species valued for their wood. Distribution: Tropical America.

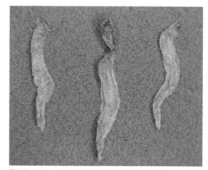

Rubiaceae - *Calycophyllum* (8.25×; 5.86 × 0.94)

Capirona Spruce. Trees to 20 m tall. Trunk very smooth and green after defoliation of the brown papyraceous bark. Leaves and axillary stipules very large. Infructescence terminal. Fruit a bivalved capsule, to 1.8 cm long, brown when mature. Seeds winged, many per fruit. Species valued for their wood. The genus *Loretoa* is synonymous with *Capirona*. Distribution: Guianas and Venezuela to Peru and Bolivia.

Rubiaceae - *Capirona* (5.71×; 7.94 × 2.22)

Chomelia Jacq. Shrubs to 4 m tall, or sometimes lianas. Leaves with inconspicuous to invisible tertiary venation. Infructescence axillary. Fruit a drupe to 1.5 cm long, black when mature, the mesocarp green transparent. Seeds one per fruit. Plants with short pubescence. Many species with spines. Distribution: Tropical America, but more speciose in Africa and Asia, to 1200 m elevation.

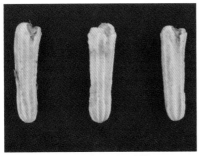

Rubiaceae - *Chomelia* (2.96×; 11.36 × 2.93)

Coussarea Aubl. Shrubs and trees to 12 m tall. Leaves sometimes coriaceous, the tertiary venation inconspicuous or invisible in some species. Stipules conspicuous, sheathing the stem. Infructescence axillary or terminal. Fruit a berry, sessile, to 2 cm long, white when mature, crowned by persistent stigma. Seeds 1–2 per fruit. Plants pubescent or glabrous. Distribution: Tropical America.

Rubiaceae - *Coussarea* (3.15×; 13.39 × 8.52)

Duroia L. f. Trees to 20 m tall. Leaves sometimes verticillate, usually pubescent. Stipules caducous, leaving a ring on the branch after falling. Infructescence terminal. Fruit a berry, to 5 cm long, green, yellow, or brown when mature, pubescent, crowned by the persistent, large, conspicuous, tubular calyx. Seeds various per fruit. Plants generally associated with ants. Distribution: Tropical South America.

Rubiaceae - *Duroia* (2.48×; 8.25 × 7.01)

Faramea Aubl. Shrubs or trees to 15 m tall. Leaves coriaceous, the secondary venation clearly anastomosing in some species. Stipules conspicuous, thin, in axillary pairs, and overlapping at the branch apex. Infructescence terminal, sometimes subtended by one yellow bract, the peduncle blue. Fruit a drupe to 1.5 cm diameter, blue or black when mature. Seeds one per fruit. Distribution: Mexico to Peru and Bolivia, to 1500 m elevation.

Rubiaceae - *Faramea* (3.32×; 9.68 × 5.73)

Rubiaceae - *Faramea* (2.79×; 7.16 × 7.01)

Ferdinandusa Pohl. Shrubs or trees to 20 m tall. Leaves short-petioled, somewhat coriaceous, rarely verticillate, the lamina glabrous or covered with fine, reddish-brown tomentose pubescence. Stipules terminal, triangular, glabrous, and caducous. Infructescence terminal. Fruit a loculicidal bivalved capsule, elongate, cylindrical, or subglobose, to 4 cm long, somewhat woody. Seeds winged, many per fruit. Shoots and bark of some species reportedly medicinal. Distribution: Panama to Guianas and Brazil, Bolivia, and Peru, to 1500 m elevation.

Rubiaceae - *Ferdinandusa* (3.03x; 8.46 × 2.2)

Genipa L. Trees to 30 m tall. Leaves large, clustered at branch apices. Infructescence axillary or terminal. Fruit a berry, to 11 cm diameter, yellow when mature, crowned by the conspicuous, persistent, elongate, tubular calyx; the mesocarp cream-colored, fragrant, oxidizing blue. Seeds many per fruit. Some species valued for their wood. The fruit edible. Mesocarp used as source of blue fabric dye and body paint. Distribution: Mexico, Cuba, Puerto Rico, Antilles, Trinidad and Tobago, to Argentina.

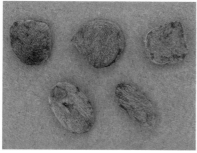

Rubiaceae - *Genipa* (2.22x; 8.71 × 7.29)

Geophila D. Don. Sprawling herbs. Leaves usually cordate at the base. Infructescence terminal. Fruit a berry, to 0.8 cm diameter, orange, red, or black when mature, crowned by persistent stigma; the mesocarp fleshy, transparent. Seeds two per fruit. Plants glabrous or pubescent. Distribution: Mexico to Peru and Bolivia.

Rubiaceae - *Geophila* (8.1x; 5.58 × 2.99)

Rubiaceae - *Geophila* (10.95x; 3.9 × 2.4)

Gonzalagunia R. & P. Shrubs to 4 m tall. Leaves sometimes asperous on abaxial laminar surface. Infructescence terminal. Fruit a berry, to 0.8 cm diameter, generally white or sometimes dark purple to black when mature. Seeds four per fruit. Plants pubescent. Distribution: Guianas and Surinam to Peru and Bolivia.

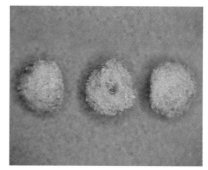

Rubiaceae - *Gonzalagunia* (9.52x; 2.03 × 1.81)

Hamelia Jacq. Shrubs to 3 m tall. Leaves petiolate, very thin, sometimes verticillate in threes. Infructescence terminal. Fruit a berry to 1 cm diameter, sessile or short-pedicillate, red to black when mature; the apex short-tubular due to persistent calyx. Seeds many per fruit. Distribution: Mexico to Peru and Bolivia.

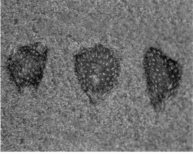

Rubiaceae - *Hamelia* (21.8x; 0.99 × 0.57)

Hillia Jacq. Epiphytes. Branches red at young apex. Leaves succulent, the venation inconspicuous below, the petiole canaliculate, short, swollen. Infructescence terminal. Fruit a septicidal capsule, to 10 cm long, striate, conspicuously crowned at the apex by a persistent calyx; brown when mature. Seeds plumose, many per fruit. Distribution: Guianas to Peru.

Rubiaceae - *Hillia* (4.39x; 2.02 × 0.67)

Ixora L. Trees to 8 m tall. Stems triangular in cross section. Leaves coriaceous, sometimes three-verticillate. Stipules completely sheathing the stem at each node. Infructescence terminal or axillary. Fruit a berry, sessile or short-pedicellate, to 1.2 cm long, yellow or orange when immature, red or dark purple when mature; the mesocarp yellowish. Seeds 1–2 per fruit. Plants with short pubescence. Distribution: Tropical America, Asia, Africa, and Australia.

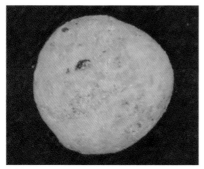

Rubiaceae - *Ixora* (9.44x; 5.12 × 4.9)

Kotchubaea Fish ex DC. Trees to 6 m tall. Leaves clustered at branch apices. Fruit to 7.3 cm long, the calyx persisting at the apex, elongate and conspicuous. Seeds many per fruit. Distribution: South America.

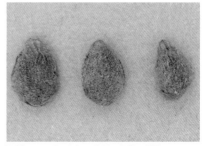

Rubiaceae - *Kotchubaea* (2.05x; 11.39 × 7.89)

Ladenbergia Klotzsch. Shrubs or trees to 20 m tall. Leaves relatively large, petiolate. Stipules conspicuous, caducous. Infructescence terminal. Fruit a septicidal capsule, to 1.2 cm long, brown when mature. Seeds winged, many per fruit. Distribution: Costa Rica to Peru and Bolivia, to 1200 m elevation.

Rubiaceae - *Ladenbergia* (3.6x; 17.15 × 3.24)

Macrocnemum P. Browne. Trees to 25 m tall. Trunks fenestrate. Leaves large, somewhat clustered at the branch apex. Stipules caducous. Infructescence terminal. Fruit a dry capsule, to 1 cm long, brown when mature. Seeds tiny, winged, many per fruit. Distribution: Costa Rica to Peru and Bolivia.

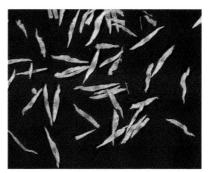

Rubiaceae - *Macrocnemum* (4.25x; 3.36 × 0.57)

Malanea Aubl. Usually lianas; some species shrubs to 5 m tall. Leaves glaucous abaxially, glabrous adaxially. Stipules conspicuous. Infructescence axillary. Fruit a drupe, sessile, curved, to 1.4 cm long, dark purple when mature, crowned by persistent calyx. Seeds one per fruit. Pubescence very dense in some species. Distribution: Guianas to Peru.

Rubiaceae - *Malanea* (3.22×; 11.97 × 2.93)

Manettia Mutis ex L. Lianas. Infructescence axillary. Fruit a septicidal capsule, bivalved, to 1.5 cm diameter, crowned by persistent calyx. Seeds many per fruit. Distribution: Mexico, Cuba, and Jamaica, to Peru.

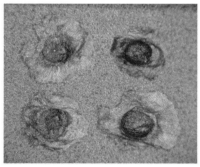

Rubiaceae - *Manettia* (11.69×; 2.27 × 1.96)

Morinda L. Shrubs to 2 m tall. Leaves lightly asperous on adaxial surface, very thin when dry. Infructescence axillary, sessile. Fruit a berry to 1 cm long,

Rubiaceae - *Morinda* (8.89×; 5.17 × 2.74)

green to yellow when mature. Seeds three per fruit. Plants with very short pubescence throughout. Distribution: Pantropical.

Oldenlandia L. Herbs to 0.4 m tall. Leaves sessile, narrow. Infructescence axillary. Fruit a capsule, to 0.3 cm long, brown when mature, 1–4 per axil. Seeds many per fruit. Distribution: Tropical America, introduced from the Old World.

Rubiaceae - *Oldenlandia* (158.73×; 0.32 × 0.22)

Palicourea Aubl. Shrubs to 8 m tall. Leaves sometimes three-verticillate, and with two pairs of persistent, axillary stipules. Infructescence terminal, the peduncle red or yellow. Fruit a berry, sessile, to 1.4 cm diameter, violet, purple to black when mature; subtended by conspicuous bracts in some species. Seeds two per fruit. Plants pubescent or glabrous. Distribution: Mexico to Peru and Bolivia, to 2300 m elevation.

Rubiaceae - *Palicourea* (9.31×; 4.51 × 2.99)

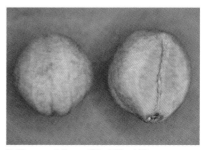

Rubiaceae - *Palicourea* (6.3×; 5.08 × 4.62)

Rubiaceae - *Palicourea* (3.98×; 7.1 × 6.22)

Pentagonia Benth. Shrubs or trees to 10 m tall, mostly unbranched. Leaves usually large, clustered at the apex of the plant. Stipules reddish. Infructescence sessile, axillary or borne from the trunk. Fruit a berry, sessile or short-pedicellate, to 3 cm diameter, brown to brownish-yellow when mature, crowned by persistent, elongate, tubular calyx; the mesocarp white to white-creamish, transparent. Seeds up to 40 per fruit. Plants with short pubescence. Distribution: Panama to Peru and Bolivia.

Rubiaceae - *Pentagonia* (7.04×; 4.12 × 3.53)

Posqueria Aubl. Shrubs or trees to 10 m tall. Leaves large, coriaceous. Infructescence terminal. Fruit a berry to 5 cm diameter, green or yellow when immature, orange when mature, crowned by the short persistent tubular calyx; the mesocarp orange. Seeds many per fruit. Distribution: Tropical America.

Rubiaceae - *Posqueria* (2.79×; 9.09 × 7.88)

Psychotria L. Usually shrubs or trees to 15 m tall, rarely epiphytes. Leaves highly variable. Infructescence terminal or sometimes axillary; some species with reddish bracts subtending the infructescence and the fruits, sometimes very conspicuous, sometimes with red or purple peduncles. Fruit a sessile berry to 1.8 cm diameter, yellow or orange when immature, white, blue, violet, red, purple, or black when mature, crowned by persistent stigma and calyx; the mesocarp white, mealy in some species. Seeds 1–2 or rarely five per fruit. Plants pubescent or glabrous. Some species with showy green or red bracts surrounding the fruits. Distribution: Mexico to Uruguay, also in the Old World, to 2000 m elevation.

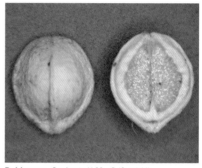

Rubiaceae - *Psychotria* (6.08×; 5.43 × 4.66)

Rubiaceae - *Psychotria* (9.1×; 6.22 × 5.03)

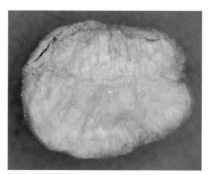

Rubiaceae - *Psychotria* (9.58×; 6.07 × 4.82)

Rubiaceae - *Psychotria* (6.46x; 6.79 × 2.97)

Rubiaceae - *Psychotria* (5.4x; 7.02 × 2.98)

Rubiaceae - *Psychotria* (8.89x; 4.37 × 3.18)

Rubiaceae - *Psychotria* (6.43x; 5.28 × 4.42)

Randia L. Shrubs or trees to 15 m tall, rarely lianas. Leaves sometimes clustered at branch apex. Infructescence axillary or borne from the trunk. Fruit a berry, to 9 cm long, yellow when mature, crowned by persistent, tubular calyx; the exocarp smooth or wrinkled; the mesocarp thin; the endocarp sometimes strong, woody. Seeds many per fruit, embedded in a black, fragrant pulp. Many species with spines on the branches. Plants pubescent or glabrous. Distribution: Southern United States and Mexico to Argentina, also in the Old World, to 3300 m elevation.

Rubiaceae - *Psychotria* (7.99x; 3.34 × 2.15)

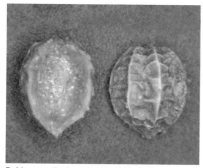

Rubiaceae - *Psychotria* (11.16x; 3.18 × 2.43)

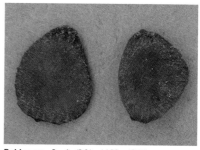

Rubiaceae - *Randia* (2.01x; 16.58 × 13.16)

Rudgea Salisb. Shrubs to 5 m tall. Leaves weakly coriaceous, rarely sessile or short-petiolate. Infructescence usually terminal, rarely axillary, sessile or pedunculate; the peduncle sometimes purple, flattened to nearly triangular. Fruit a berry, to 1.5 cm diameter, white when mature, crowned by the persistent calyx;

the mesocarp white, mealy. Seeds 1–2 per fruit. Plants with short pubescence. Distribution: Mexico to Peru and Bolivia.

Rubiaceae - *Rudgea* (6.22×; 5.83 × 4.56)

Sabicea Aubl. Lianas. Stipules in one pair oriented downward. Infructescence axillary. Fruit a berry, to 1 cm diameter, purple, blue, or black when mature, crowned by persistent calyx. Seeds tiny, many per fruit. Plants with dense reddish or yellowish pubescence throughout. Common in disturbed areas. Distribution: Mexico to Peru and Bolivia, but more speciose in Africa.

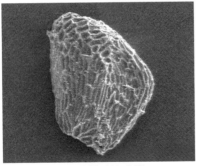

Rubiaceae - *Sabicea* (79.37×; 0.61 × 0.45)

Simira Aubl. Trees to 15 m tall. Infructescence terminal. Fruit a 4-valved capsule, to 3.5 cm diameter, brown when mature. Seeds winged, various per fruit, sometimes with white mottling. Distribution: Mexico to Peru and Bolivia.

Rubiaceae - *Simira* (1.88×; 31.12 × 10.96)

Rubiaceae - *Simira* (2.42×; 19.74 × 10.48)

Uncaria Sche. Lianas. Stems with one pair of curved or straight spines at each node. Infructescence in the form of a globose head, axillary or terminal. Fruit a 4-valved capsule, to 2 cm long, brown when mature. Seeds tiny, winged, many per fruit. Common in disturbed areas. Distribution: Central America to Peru and Bolivia, but more speciose in the Old World.

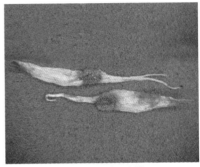

Rubiaceae - *Uncaria* (6.77×; 8.95 × 1.02)

Warscewiczia Klutsch. Trees to 16 m tall. Stipules caducous. Infructescence terminal or axillary, subtended by a persistent red lobe of the calyx. Fruit a septicidal capsule, to 1 cm long, brown when mature. Seeds tiny, winged, many per fruit. Distribution: Mexico to Peru and Bolivia, to 1200 m elevation.

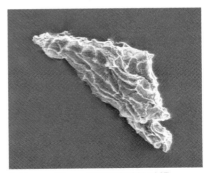

Rubiaceae - *Warscewiczia* (62.43×; 0.96 × 0.37)

RUTACEAE

Trees and shrubs, rarely herbs. Leaves simple, or pinnately, bipinnately, or palmately compound, generally alternate, rarely opposite, with translucent punctations, especially along margins, and with citric fragrance in most species when crushed. Infructescence axillary or terminal. Fruit a follicle or capsule, berry or hesperidium, or samara. Many genera with spines on the trunk, stems, and leaves.

Esenbeckia H. B. K. Shrubs and trees to 6 m tall. Leaves simple or compound with 3–5 leaflets, alternate, the translucent punctations inconspicuous; the petioles pulvinate and variable in length. Infructescence axillary or terminal. Fruit a 5-valved woody capsule, to 3 cm long, brown when mature, smooth or with prickles. Seeds one per locule, up to 10 per fruit. Plants pubescent or glabrous. Distribution: Mexico to Argentina.

Rutaceae - *Esenbeckia* (5.67x; 10.6 × 5.56)

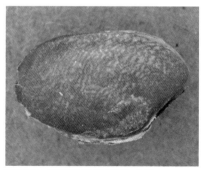

Rutaceae - *Esenbeckia* (5.65x; 10.49 × 6.67)

Galipea Aublet. Shrubs or trees to 15 m tall. Leaves alternate or opposite, simple or trifoliately compound. Infructescence axillary or terminal. Fruit a 5-valved capsule, to 3 cm long, subwoody, brown when mature. Seeds one per valve of the fruit, or up to 10 per fruit. Distribution: Guatemala and the West Indies to Bolivia and Peru, to 1100 m elevation.

Rutaceae - *Galipea* (2.1x; 9.34 × 8.33)

Zanthoxylum L. Shrubs or trees to 35 m tall. Trunks of most species with small to large thorns. Leaves compound, paripinnate or imparipinnate, with up to 12 opposite leaflets; strongly aromatic when crushed; the translucent punctations conspicuous or inconspicuous. Infructescence terminal or borne from the stems. Fruit a bivalved capsule, to 1.5 cm long, brown when mature. Seeds one per fruit. Wood used for lumber. Common in disturbed areas. Distribution: Pantropical, to 2500 m elevation.

Rutaceae - *Zanthoxylum* (8.47x; 3.95 × 3.04)

SABIACEAE

Trees, rarely shrubs. Leaves simple or pinnately compound, alternate. Infructescences axillary or terminal. Fruit a drupe.

Meliosma Blume. Trees to 20 m tall. Leaves simple, the margins entire, the secondary venation weakly anastomosing throughout and strongly reticulate abaxially; the petiole base swollen. Infructescences terminal. Fruit a drupe, to 2 cm diameter, yellow when mature, the pedicel base inserted to one side. Seeds one per fruit. Distribution: Tropical America, Asia, to 3300 m elevation.

Sabiaceae - *Meliosma* (2.03×; 14.58 × 13.59)

Ophiocaryon Endl. Trees to 30 m tall. Leaves compound, paripinnate or imparipinnate, the leaflets opposite or subopposite. Infructescence terminal. Fruit a drupe to 3.5 cm diameter, somewhat asymmetric, yellow-green when mature. Seed one per fruit. Distribution: Brazil and northeastern Peru.

Sabiaceae - *Ophiocaryon* (1.32×; 25.92 × 20.88)

SALICACEAE

Trees. Leaves simple, alternate, entire, rarely lobed. Stipules present. Of the two species in the Neotropical region, only one is native. Many species introduced for their wood and as ornamentals.

Salix L. Trees to 20 m tall. Leaves linear to lanceolate, the margins serrate. Stipules persistent or caducous. Infructescence axillary. Fruit a bivalved capsule, sessile, to 0.6 cm long. Many tiny seeds per fruit, surrounded by white, cotton-like fibers. Wood used

for lumber. The bark is rich in tannins and is the source of salicylic acid, the base of commercial aspirin. Bark and roots medicinal. Very common along river banks. Distribution: Tropical and subtropical America, from the southern United States and Mexico to Chile and Argentina; to 3000 m elevation; most species in temperate zones.

Salicaceae - *Salix* (4.76×; 0.8 × 0.33)

SAPINDACEAE

Lianas, shrubs, and trees. Leaves compound, unifoliate, trifoliate, pinnate, rarely bipinnate, always alternate; the raquis sometimes winged; in tree genera, the pinnately compound leaf terminating in a small aborted bud arising from one side of the rachis, and the leaflets alternate or subopposite. Some genera of lianas with latex. Infructescence axillary, terminal, or borne from the trunk. Fruit a capsule, drupe, or samara. In genera with capsules, the seeds winged. Plants glabrous or pubescent. Aerial roots and triangular stems common in lianas.

Allophylus L. Shrubs or trees to 10 m tall. Leaves trifoliate, rarely unifoliate, the leaflet margins entire or serrate along the upper half. Infructescence axillary, terminal, or borne from the stems. Fruit druplike, to 1.5 cm long, yellow or orange when immature, red when mature, subtended by persistent calyx; the mesocarp yellow or red, fleshy. Seeds one per fruit. Plants pubescent or glabrescent. Distribution: Pantropical, to 2000 m elevation.

Salicaceae - *Salix* (30.79×; 0.96 × 0.42)

Sapindaceae - *Allophylus* (3.68×; 10.78 × 7.47)

Averrhoideum Baill. Trees to 25 m tall, sometimes shrubs. Leaves compound, usually paripinnate, always terminating in an aborted bud; the leaflets alternate or subopposite; the margins entire or serrate at the apex. Infructescence terminal or axillary. Fruit a bivalved capsule, to 2.5 cm long, asymmetric, dehiscing irregularly from the base to the apex, red when mature. Seeds one per fruit, arillate. Distribution: Mexico to Brazil and Paraguay, to 500 m elevation.

Sapindaceae - *Averrhoedeum* (3.22x; 7.6 × 6.84)

Cardiospermum L. Herbaceous vines with tendrils. Leaves pinnately compound with many leaflets. Fruit a 3-celled inflated capsule, membranous, nerved, to 5 cm long, brown when mature. Seeds one per valve of the fruit. Some species reportedly used as aphrodisiacs or medicine. Distribution: Southern United States to lowland Amazonia, also in Asia and Africa.

Sapindaceae - *Cardiospermum* (2.52x; 23.74 × 11.26)

Cupania L. Trees to 20 m tall. Leaves compound, imparipinnate, glabrous or pubescent; the rachis terminating in a conspicuous bud; the leaflet margins entire, crenate, or dentate. Infructescence terminal or axillary. Fruit a 2- to 5-valved capsule, round, 1-2 cm diameter, reddish when mature, with some of the capsular valves aborting. Seeds 1–2 per fruit, with a orange or yellow aril. Distribution: Mexico to Peru and Bolivia.

Sapindaceae - *Cupania* (1.76x; 10.9 × 7.89)

Matayba Aubl. Trees to 20 m tall. Leaves compound, imparipinnate, with 2–10 pairs of alternate or subopposite leaflets; the secondary venation conspicuous, reticulate. Infructescence axillary or terminal. Fruit a loculicidal capsule, bi- to trivalved, to 2 cm long, asymmetric, sometimes apiculate, red when mature. Seeds 1–2 per fruit, covered completely by a white aril. Plants glabrous. Distribution: Trinidad, St. Vincent, Guianas, Bolivia, and Peru.

Sapindaceae - *Matayba* (3.58x; 9.97 × 7.6)

Sapindaceae - *Matayba* (5.4x; 10 × 7.59)

Paullinia L. Generally lianas, rarely shrubs to 2 m tall. Stems sometimes 3- to 4-angled or striate. Tendrils axillary when lianas. White latex in many species. Leaves compound, imparipinnate, with 3–7 opposite leaflets, or ternate/trifoliate in some species; the margins weakly serrate to dentate; the raquis sometimes broadly winged. Infructescence with one to various fruits, axillary or borne from the trunk. Fruit a septicidal capsule, to 3.5 cm diameter, smooth or with long spine-like projections, sometimes three-winged; yellow, orange, or red when mature, and subtended by persistent calyx. Seeds 1–3 per fruit, covered one-third to completely by a white aril. Plants pubescent or glabrous. One species widely cultivated for seeds used in the production of the caffeinated soft drink *guarana* . Distribution: Mexico to Brazil, Bolivia, and Peru; also in Africa and Madagascar.

Sapindaceae - *Paullinia* (2.75×; 16.58 × 8.5)

Sapindaceae - *Paullinia* (4.49×; 10.5 × 9.13)

Sapindaceae - *Paullinia* (4.04×; 10.84 × 7.28)

Sapindus L. Trees to 30 m tall. Leaves compound, imparipinnate, the leaflets opposite or subopposite, the raquis weakly winged. Infructescence terminal. Fruit a capsule, to 1.5 cm diameter, red when mature; the exocarp subwoody, with one or two fused, atrophic carpels, giving a peculiar aspect to the fruit, such as the form of ears. Seeds one per fruit. Distribution: Tropical America and the Old World.

Sapindaceae - *Sapindus* (1.79×; 18.93 × 16.86)

Serjania Mill. Lianas. Stems of some species triangular or with white latex. Tendrils usually not branched. Leaves compound, usually biternate, or pinnate with five leaflets; the margins weakly dentate on the upper half; the adaxial laminar surface glossy. Infructescence axillary. Fruits composed of three samaras longitudinally fused at the end opposite the seed; each samara to 4 cm long, red when immature, brown when mature; subtended by persistent calyx. Seeds one per samara. Plants pubescent or glabrous. Distribution: Southern United States, Cuba, Mexico, to Peru and Bolivia.

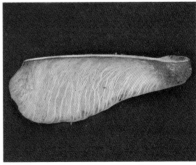

Sapindaceae - *Serjania* (2.65×; 22.96 × 8.36)

Talisia Aubl. Shrubs or trees to 25 m tall, frequently lacking branches. Leaves clustered at the trunk apex, compound, imparipinnate, with up to 25 leaflets, opposite or alternate; the leaflet base asymmetric; the venation weakly anastomosing; the adaxial leaflet surface glossy; the petiole base swollen; the raquis sometimes more than 1 m long, terminating in an aborted bud attached to the apical leaflet. Infructescence axillary or terminal, subtended by a spine-like bracteole. Fruit a drupe, to 3 cm long, apiculate, yellow or orange when mature, subtended by persistent calyx; the exocarp subwoody; the mesocarp white, fleshy. Seeds one per fruit. Distribution: Tropical America.

Sapindaceae - *Talisia* (2.18×; 20.63 × 12.67)

Sapindaceae - *Talisia* (2.52×; 18.82 × 9.16)

Thinouia Triana & Planch. Lianas. Tendrils axillary or epipeduncular, two-branched. Leaves trifoliately compound with translucent punctations. Stipules absent or reduced. Infructescence axillary or terminal. Fruit composed of three samaras fused at the base,

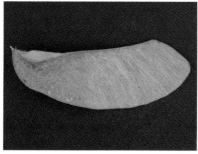

Sapindaceae - *Thinouia* (1.54×; 39.11 × 13.29)

each winged on the distal plane, to 5 cm long. Seeds one per samara. Distribution: South America.

Toulicia Aubl. Trees to 20 m tall. Leaves compound, imparipinnate, with up to 12 alternate or opposite, coriaceous, subsessile leaflets; the raquis terminating in an aborted bud attached to the apical leaflet. Infructescence terminal. Fruit composed of three samaras longitudinally fused at the end opposite the seeds, each samara to 3.5 cm long, brown when mature. Seeds one per samara. Distribution: Tropical America.

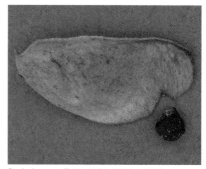

Sapindaceae - *Toulicia* (1.66×; 35.41 × 16.69)

Urvillea H.B.K. Lianas. Leaves trifoliately compound, the base swollen, the leaflet margins dentate-serrate. Infructescence axillary. Fruit composed of three samaras fused longitudinally, each samara to 3 cm long, inflated, brown when mature. Seeds one per samara. Plants with dense, short pubescence throughout. Distribution: Mexico to Paraguay and Argentina.

SAPOTACEAE

Trees. Stipules usually present, sometimes caducous. Leaves simple, alternate, entire, sometimes grouped at the apex of branches. Latex white, rarely reddish, present in small quantities. Infructescence axillary or borne from the stems. Fruit a berry. The presence of a longitudinal scar along one side of the seed allows easy identification of the family. Plants often pubescent, the hairs T-shaped.

Chrysophyllum L. Trees to 35 m tall, rarely shrubs. Latex sometimes present, reddish in color. Leaves sometimes clustered at the branches apices; the secondary venation strongly reticulate; the tertiary venation very fine and parallel. Fruit short, pedicellate or sessile, to 8 cm diameter, yellow, orange, red, or dark purple when mature; glabrous or pubescent; the mesocarp white. Seeds 1–5 per fruit. Plants weakly pubescent, the hairs bifid or T-shaped. Some species used as a source of lumber. Others cultivated for edible fruit. Distribution: Mexico to Argentina, to 1700 m elevation.

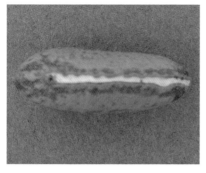

Sapotaceae - *Chrysophyllum* (2.75x; 21.96 × 7.73)

Sapotaceae - *Chrysophyllum* (2.88x; 21.29 × 8.68)

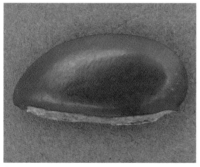

Sapotaceae - *Chrysophyllum* (2.35x; 25.36 × 12.84)

Sapotaceae - *Chrysophyllum* (4.36x; 26.14 × 7.06)

Manilkara Adans. Trees to 35 m tall. Trunks with fissured bark. Latex abundant in some species, white in color. Leaves grouped at the extremes of the branches; the secondary venation very fine, inconspicuous, and nearly perpendicular to the midvein. Stipules sometimes present, typically caducous. Infructescence axillary. Fruit 2.5 cm diameter, rarely to 10 cm diameter, orange, red or black when mature; the mesocarp white. Seeds generally one per fruit, rarely two or more. Plants glabrous. Species valued for their hard, resistant lumber that is often used for main posts and columns in building structures. The latex of one species, *Manilkara zapota*, is used to produce chewing gum. Distribution: Mexico to Paraguay, also in the Old World.

Sapotaceae - *Manilkara* (3.83x; 15.69 × 6.91)

Pouteria Aubl. Trees to 30 m tall. Trunks usually with buttress roots, the bark fissured, sometimes defoliating in small plates. Stems often angled. Leaves highly variable, often clustered at branch apices, sometimes appearing opposite; the base generally decurrent; the petiole usually swollen at base, taking the form of a bottle; the venation usually reticulate, sometimes anastomosing. Infructescence axillary or borne from the stems. Fruit solitary or in groups, sessile or pedicellate, to 10 cm diameter (rarely to 20 cm diameter), with or without an apicular projection; yellow, orange, or brown when mature, smooth or rugose, weakly tomentose or glabrous. Plants glabrous, lightly pubescent, or tomentose. Wood used for lumber. The latex of some species used industrially. Some species cultivated for edible fruit. Distribution: Mexico and Guianas to the Atlantic Coast of Brazil and to Chile, also in Asia, to 3000 m elevation.

Sapotaceae - *Pouteria* (1.59x; 28.13 × 13.2)

Sapotaceae - *Pouteria* (2.24×; 20.42 × 11.79)

Sapotaceae - *Pouteria* (1.26×; 35.29 × 20.17)

Sarcaulus Radlk. Trees to 10 m tall. Leaves relatively shorter than most other Sapotaceae genera, the venation anastomosing. Fruit a berry, to 2.5 cm diameter, yellow when mature; the mesocarp white. Seeds 1–5 per fruit. Distribution: Brazil, Peru, and Bolivia.

Sapotaceae - *Pouteria* (2.35×; 20 × 9.64)

Sapotaceae - *Sarcaulus* (3.85×; 15.38 × 6.87)

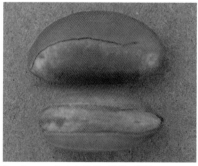

Sapotaceae - *Pouteria* (2.69×; 18.07 × 9.25)

Sapotaceae - *Sarcaulus* (3.89×; 15.38 × 9.1)

SCROPHULARIACEAE

Herbs, erect or prostrate, rarely lianas; one species can be a shrub to 2 m tall. Stems often square. Leaves simple, opposite, sometimes alternate or verticillate, sessile or petiolate; the margins entire, serrate, or crenate. Infructescence axillary. Fruit a capsule or berry.

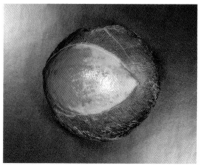

Sapotaceae - *Pouteria* (1.15×; 35.78 × 33.3)

Scoparia L. Herbs to 0.8 m tall. Leaves three-verticillate, the base very decurrent, sometimes to stem, the margin deeply serrate on the upper half. Fruit a capsule, to 0.3 cm diameter, crowned by persistent stigma. Seeds many per fruit. Distribution: Pantropical, growing in disturbed areas.

Scrophulariaceae - *Scoparia* (100.53x; 0.41 × 0.26)

SIMAROUBACEAE

Shrubs and trees. Leaves compound, imparipinnate, rarely simple, always alternate; the leaflets alternate, subopposite, or opposite; the raquis sometimes winged. Infructescence terminal or axillary. Fruit a berry or drupe, rarely samara.

Picramnia Sw. Trees to 15 m tall. Leaves compound, imparipinnate, the leaflets 5–19, alternate or subopposite, weakly asymmetric at the base; the lower leaflets completely rounded or smaller; the adaxial laminar surface glossy; the midvein sometimes positioned toward one side. Infructescence axillary. Fruit a berry, to 1.5 cm long, yellow when immature, orange, red, to dark purple when mature, crowned by persistent stigma. Seeds 1–2 per fruit. Plants glabrous or pubescent. Often segregated into the family Picramniaceae. Distribution: Tropical and Subtropical America, to 2200 m elevation.

Simaroubaceae - *Picramnia* (6.39x; 9.19 × 4.87)

Simaba Aubl. Shrub or trees to 30 m tall. Leaves compound, imparipinnate, the leaflets 1–13, opposite, the base asymmetric, the secondary venation sometimes inconspicuous. Infructescence terminal or axillary. Fruit a drupe to 3 cm long, yellow or orange when mature; the mesocarp orange. Seeds one per fruit. Plants with dense pubescence. Distribution: Trinidad to Brazil, Peru, and Bolivia.

Simaroubaceae - *Simaba* (1.97x; 21.94 × 12.1)

Simarouba Aubl. Trees to 30 m tall. Leaves somewhat grouped at the branch apices, compound, imparipinnate, the leaflets up to 20, alternate or subopposite, coriaceous; the venation inconspicuous or the secondary venation sometimes conspicuous above; the base weakly asymmetric. Infructescence terminal, composed of groups of 1–3 short-stipitate fruits on a receptacle. Fruit a drupe, to 1.6 cm long, black when mature; the mesocarp green. Seeds one per fruit. Wood used as lumber and in the preparation of paper pulp. Distribution: Central America, Trinidad and Tobago, to Brazil and Bolivia.

Simaroubaceae - *Simarouba* (3.2x; 13.08 × 7.85)

SIPARUNACEAE

Shrubs and trees, strongly aromatic. Leaves simple, opposite, sometimes verticillate. Infructescence axillary. Fruit technically an aggregate composed of many simple drupelets enclosed in the swollen hypanthium. Family segregated from the Monimiaceae.

Siparuna Aubl. Shrubs or trees to 25 m tall. Leaves opposite or three-verticillate, the margins dentate, the adaxial laminar surface sometimes asperous. Infructescence axillary with 1–2 aggregate fruits per axil. Fruit to 4 cm diameter, pink or yellow when mature, lenticellate, often strongly aromatic, dehiscing completely or irregularly when mature to expose a white or pink, somewhat mealy mesocarp. Seeds 4–20 per fruit, sometimes with red aril. Plants usually with pubescence of compound hairs, rarely glabrescent. Majority of species with white scales throughout the plant, often more dense on the branch apex and new growth. Distribution: Southern Mexico through Central America and Guianas to Peru and Bolivia.

Siparunaceae - *Siparuna* (6.61x; 5.57 × 4.45)

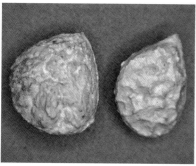

Siparunaceae - *Siparuna* (5.98x; 5.72 × 5.59)

Siparunaceae - *Siparuna* (8.04x; 3.97 × 3.59)

Siparunaceae - *Siparuna* (7.72x; 4.33 × 3.74)

SMILACACEAE

Lianas. Stems green and generally with spines. Tendrils paired, arising from the petiole. Leaves simple, alternate, with 3–7 parallel secondary veins; the petiole base nearly sheathing the stem. Infructescence axillary or terminal. Fruit a berry. In some herbaria and old texts the members of this family are included within the Liliaceae.

Smilax L. Lianas. Tendrils present. Petiole elongate, perpendicular to the stem, generally rolled when dry. Infructescence axillary. Fruit to 1.7 cm diameter, orange, red, to black when mature; the mesocarp orange; the endocarp transparent and elastic, surrounding the seed. Seeds 1–4 per fruit, round when one seed or with flattened sides when more than one seed. The stem and infructescence with small curved or erect spines. Plants pubescent or glabrous. Distribution: Pantropical, and also in temperate zones.

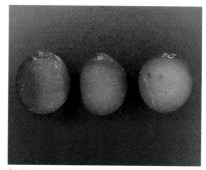

Smilacaceae - *Smilax* (2.73x; 7.91 × 6.94)

SOLANACEAE

Herbs and shrubs, also lianas, hemiepiphytes, and trees. Leaves simple or compound, alternate (rarely opposite but the leaves of different sizes); the margins entire or lobed. Infructescence axillary or terminal. Fruit a berry or capsule, subtended by persistent, often conspicuous calyx. The majority of genera with a fetid odor when crushed. Plants glabrous or pubescent, the hairs simple or stellate.

Brunfelsia L. Shrubs to 5 m tall. Leaves simple, very thin and nearly translucent when dry, but sometimes coriaceous. Infructescence terminal or axillary. Fruit a berry, to 3 cm diameter, green-yellowish when mature, subtended by persistent calyx. Seeds many per fruit. Plants glabrous or pubescent. Some species cultivated as ornamentals. Distribution: Tropical America, to 3200 m elevation.

Solanaceae - *Brunfelsia* (7.51x; 3.87 × 2.82)

Capsicum L. Herbs, sometimes climbing or scrambling, rarely shrubs to 2 m tall. Stems smooth, yellow, weakly striate, square, hollow. Leaves simple, or when opposite with one leaf in the pair smaller than the other, very thin when dry. Infructescence axillary, terminal, or borne from the stems, composed of 2–8 fruits. Calyx weakly lobed, with apices facing downward, sometimes reduced. Fruit a berry, to 2.8 cm diameter, red, orange, or yellow, and with hot, spicy flavor when mature; subtended by persistent, weakly-lobed calyx, the lobes facing downward. Some species cultivated and domesticated for the fruits used as a spice. Distribution: Florida and southern United States to Argentina, to 3400 m elevation.

Solanaceae - *Capsicum* (7.14x; 4.09 × 2.65)

Cestrum L. Shrubs or trees to 10 m tall (rarely more than 20 m tall), or lianas. Leaves simple, very thin in some species. Infructescence axillary. Fruit a berry to 2 cm long, dark purple to black when mature, sur-

rounded by persistent calyx at the base; the mesocarp white. Seeds five per fruit. Plants glabrous or pubescent, the hairs simple or branched. Distribution: Tropical and subtropical America, to 3800 m elevation.

Solanaceae - *Cestrum* (5.45x; 5.24 × 3.42)

Solanaceae - *Cestrum* (5.77x; 6.64 × 3.43)

Cyphomandra Sendtner. Shrubs or trees to 10 m tall. Leaves simple, thin, nearly translucent when dry, the margins entire or lobed, the base cordate and asymmetric. Infructescence axillary, descending. Fruit a berry to 8 cm diameter, green, with light green longitudinal lines and a strong odor when mature. Seeds many per fruit. Plants glabrous or pubescent, with a characteristic dichotomous branching pattern. Distribution: Honduras to Peru, Brazil, and Bolivia, to 2800 m elevation. Included in the genus *Solanum*.

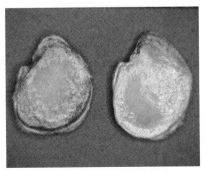

Solanaceae - *Cyphomandra* (4.81x; 7.52 × 5.76)

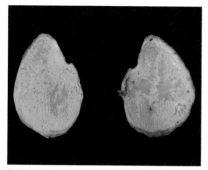

Solanaceae - *Cyphomandra* (4.53x; 7.73 × 5.82)

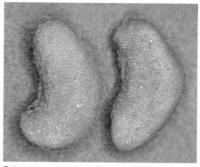

Solanaceae - *Juanulloa* (9.79x; 4.45 × 2.62)

Lycianthes (Dunal) Hassler. Herbs, lianas, or shrubs to 4 m tall. Leaves simple, the margins entire, the base decurrent, asymmetric or cordate; some species with one smaller leaf opposite the normal. Infructescence axillary. Fruit a berry, to 2 cm diameter, orange or red when mature, subtended and sometimes covered to one-third length by persistent calyx lobes that point downward. Seeds four to many per fruit. Plants glabrous or pubescent, the hairs simple or stellate. Distribution: Mexico to Paraguay, also in Asia, to 3800 m elevation.

Solanaceae - *Cyphomandra* (3.47x; 6.4 × 5.12)

Hawkesiophyton Hunz. Epiphytes growing in ant gardens. Leaves simple, membranous or fleshy-coriaceous, the base strongly decurrent, the tertiary venation inconspicuous. Infructescence terminal. Fruit to 1.5 cm diameter, green when mature, covered completely by lobed calyx. Seeds many per fruit. Branches sometimes with adventitious rooting. Related to genus *Markea*. Distribution: South America.

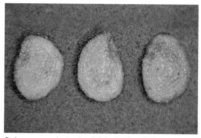

Solanaceae - *Lycianthes* (10.9x; 2.09 × 1.52)

Solanaceae - *Hawkesiophyton* (11.32x; 3.57 × 1.25)

Juanulloa R. & P. Shrubs, trees to 8 m tall, or lianas, often growing as hemiepiphytes in canopy trees. Leaves simple, coriaceous, the base decurrent, the petiole curved at the base. Infructescence terminal. Fruit a berry, to 2.5 cm diameter, the mesocarp green. Seeds 15–18 per fruit. Distribution: Peru and Bolivia, to 1500 m elevation.

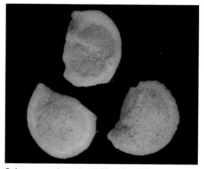

Solanaceae - *Lycianthes* (6.08x; 4.5 × 4.21)

Markea Rich. Herbs, usually epiphytes or hemiepiphytes, rarely shrubs; some species growing in ant gardens. Leaves alternate, simple, coriaceous, clustered at the branch apex, sometimes with punctations, usually glabrous; the margins entire; the tertiary venation nearly invisible. Infructescence axillary or terminal, sometimes with peduncles to 20 cm long. Fruit a berry to 2 cm long, yellow when mature, covered by the two persistent calyx lobes to 4 cm long and generally light green. Distribution: Panama, Guianas, and the Central Amazon to Peru, to 700 m elevation.

Solanaceae - *Markea* (7.64×; 6.01 × 3.2)

Nicotiana L. Herbs, shrubs, or trees to 7 m tall. Leaves simple, sometimes the base decurrent or very cordate with the lobes extending to and sheathing the stem. Infructescence axillary and terminal. Fruit a capsule, to 2 cm long, brown when mature, partially or completely covered by calyx. Seeds tiny, many per fruit. Plants glabrous or pubescent. Widely cultivated for production of tobacco. Distribution: Tropical and subtropical America, also in Africa and Australia, to 3600 m elevation.

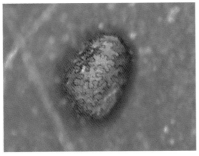

Solanaceae - *Nicotiana* (49.21×; 0.61 × 0.43)

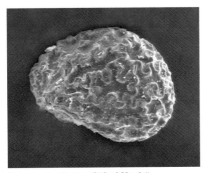

Solanaceae - *Nicotiana* (96.3×; 0.53 × 0.4)

Physalis L. Herbs to 1 m tall. Leaves alternate, variable in size throughout the plant. Infructescence axillary. Fruit a berry to 1.5 cm diameter, yellow when mature, completely covered by persistent, inflated calyx. Seeds various per fruit. Plants glabrous or pubescent. Distribution: Canada to Chile, also in the Old World.

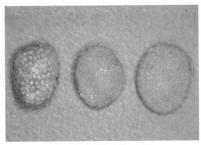

Solanaceae - *Physalis* (15.87×; 1.47 × 1.18)

Solanum L. Lianas, herbs, shrubs, or trees to 10 m tall. Leaves simple, sometimes with one smaller leaf opposite the other, often thin when dry; the base decurrent or asymmetric, the margins entire or lobed, the venation sometimes anastomosing; some species with compound leaves, imparipinnate, the leaflets opposite and the rachis sometimes winged. Infructescence opposite the leaf, axillary or borne from the stems or trunk, rarely terminal. Fruit a berry to 5 cm diameter, yellow, orange, or red, or rarely black when mature, glabrous or tomentose; subtended by persistent calyx that is lobed, entire, or in the form of a cupule; the mesocarp juicy. Seeds few to many per fruit. Plants glabrous or pubescent, the hairs simple to stellate. Some species with angled stems or curved yel-

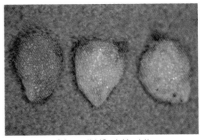

Solanaceae - *Solanum* (16.19×; 1.44 × 1.1)

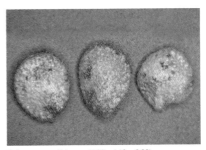

Solanaceae - *Solanum* (6.08×; 4.12 × 3.23)

low, orange, or brown spines on the stem or abaxial surface of the midvein. Many species cultivated for their edible fruits and tubers. Many species of disturbed areas. Distribution: Canada and the United States to Argentina, also in the Old World to 3500 m elevation.

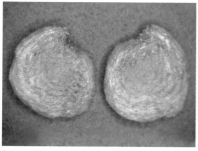
Solanaceae - *Solanum* (7.78x; 4.19 × 3.85)

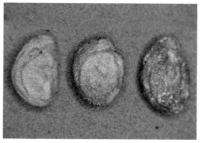
Solanaceae - *Solanum* (7.35x; 3.53 × 2.43)

STAPHYLEACEAE

Trees. Leaves compound, imparipinnate, alternate or opposite, the leaflets opposite with serrate or crenate margins. Fruit a drupe. Seeds one per fruit.

Huertea R. & P. Trees to 30 m tall. Leaves compound, alternate. Infructescence axillary. Fruit a drupe to 1.5 cm long, dark purple to black when mature, subtended by persistent calyx in the form of a cupule. Seeds one per fruit. Distribution: West Indies to Peru.

Staphylaceae - *Huertea* (1.71x; 12.84 × 11.6)

Turpinia Vent. Trees to 20 m tall. Leaves compound, opposite, the leaflets opposite. Infructescence terminal. Fruit a drupe, to 0.8 cm diameter, yellow when mature. Seeds one or two per fruit. Distribution: Tropical America and Asia, to 2000 m elevation.

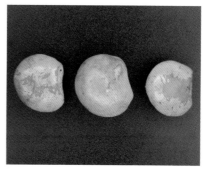
Staphylaceae - *Turpinia* (2.67x; 8.02 × 7.46)

STERCULIACEAE

Trees, sometimes shrubs or lianas. Leaves simple, alternate, entire or lobed, 3- to 5-veined, or palmately compound; the petiole conspicuously pulvinate. Stipules present, usually caducous. Infructescence axillary, terminal, or borne from the stems or trunk. Fruit a capsule or follicle, sometimes from an apocarpous gynoecium, or a berry. Seeds sometimes winged. Many species pubescent with simple or stellate hairs.

Byttneria Loefl. Lianas. Leaves cordate, 3- to 5-nerved; the base truncate or cordate; the lamina tomentose or glaucous on abaxial surface; the margins entire or serrate; the petioles variable in size and weakly swollen or bent at both extremes. Fruit a 5-valved capsule, to 6 cm diameter, brown or black when mature, spinose, densely muricate, or with caducous spine-like structures. Seeds one per carpel. Plants glabrescent to tomentose, the hairs stellate. Some species with spines or angled stems. Distribution: Mexico to Peru, also in the Old World.

Sterculiaceae - *Byttneria* (2.24x; 16.08 × 12.5)

Sterculiaceae - *Byttneria* (2.2x; 17.84 × 5.19)

Sterculiaceae - *Guazuma* (8.15x; 3.08 × 2.16)

Guazuma Mill. Trees to 28 m tall. Leaves caducous, the abaxial laminar surface densely pubescent to tomentose; the petioles with swollen pulvini at both extremes, the base weakly cordate or asymmetric; the margins irregularly dentate-crenate. Infructescence axillary. Fruit a capsule, dehiscent in five segments or not, to 3.5 cm diameter, densely covered by long hairs or conical projections, brown or black when mature. Seeds many per fruit, embedded in a white, sweet mesocarp when the fruit opens irregularly by pores, much less so when the capsule dehisces completely. Plants pubescent, the hairs stellate. Many species typical of disturbed areas. Sap from the wood used to clear the syrup in the production of sugar. Distribution: Mexico and West Indies to Argentina and Paraguay. Introduced in India, Hawaii, and Java, to 1300 m elevation.

Herrania Goudot. Shrubs or trees to 7 m tall. Leaves palmately compound, weakly grouped at the trunk apex, the leaflets sometimes with the midvein oriented to one side and the base weakly lobed. Infructescence borne from the trunk, often positioned at the base. Fruit a berry to 15 cm long with five longitudinal lines, brown or dark purple when mature; the exocarp subwoody; the mesocarp white, sweet. Seeds many per fruit. Plants pubescent, the hairs short, stellate. Distribution: Costa Rica to Peru.

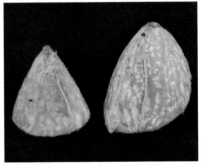

Sterculiaceae - *Herrania* (3.17x; 11.53 × 9.1)

Pterygota Schott & Endl. Trees to 40 m tall. Leaves simple, the base cordate. Infructescence axillary. Fruit a follicle to 12 cm long, brown when mature. Seeds winged, various per fruit. Distribution: Tropical America and Africa.

Sterculiaceae - *Guazuma* (0.81x; 12.08 × 7.01)

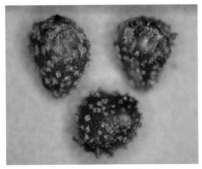

Sterculiaceae - *Guazuma* (16.61x; 1.69 × 1.3)

Sterculiaceae - *Pterygota* (0.7x; 89.7 × 30.61)

Sterculia L. Trees to 35 m tall. Leaves entire or 3- to 5-lobed, grouped at the extremes of the branches, caducous, the petioles of some species very long, generally pulvinate at both extremes. Stipules terminal. Infructescence axillary or terminal. Fruit a follicle in groups of up to five, woody or semi-woody, to 22 cm long, brown or reddish when mature, the interior walls with short urticating hairs. Seeds one to various per fruit, normally many of them unviable. Plants pubescent, the hairs stellate, very dense to tomentose. Introduced throughout the tropics. Wood valued for lumber. Seeds edible, toasted or cooked. Distribution: Mexico and West Indies to Peru and Brazil, also in the Old World.

Sterculiaceae - *Sterculia* (1.77x; 27.54 × 16.89)

Theobroma L. Trees to 15 m tall. Leaves simple, three-veined from the weakly cordate base; the margins entire; the petioles with two pulvini, the upper more swollen. Infructescence axillary and borne from the stems and trunk. Stipules thin. Fruit a berry, to 30 cm long, yellow, orange, or brown when mature, weakly 5-angled with prominent longitudinal lines; the exocarp semi-woody. Seeds up to 50 per fruit, surrounded by a very sweet, white to transparent aril. Plants glabrous or with very short pubescence, the hairs stellate. Some species with glaucous abaxial laminar surface or asymmetric leaf bases. Widely cultivated throughout the world for production of chocolate from the seeds. Distribution: Mexico to Peru and Bolivia.

Sterculiaceae - *Theobroma* (2.35x; 20.81 × 10.09)

STRELITZIACEAE

Herbs, large, tree-like. Stems rather woody. Leaves simple, petiolate, the base sheathing the stem. Infructescence terminal. Fruit a loculicidal capsule. Family segregated from Musaceae.

Phenakospermum Endl. Herbs to 10 m tall. Stem rather woody, not branched. Leaves to 4 m long, grouped at the apex of the central stem. Fruit a capsule, to 10 cm long, brown when mature, the exocarp woody, indurate. Seeds various per fruit, covered with fibrous red aril. Widely cultivated as an ornamental. Distribution: Guianas to Peru and Bolivia.

Strelitziaceae - *Phenakospermum* (with aril)
(4.02x; 14.39 × 9.61)

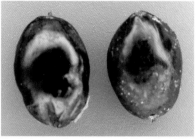

Strelitziaceae - *Phenakospermum* (6.18x; 5.87 × 4.28)

THEOPHRASTACEAE

Shrubs, sometimes small trees. Leaves simple, alternate, coriaceous, often clustered at the stem apex; the margins entire. Fruit a berry.

Clavija R. & P. Shrubs from 0.4 to 3 m tall, lacking branches. Leaves generally elongate, with strongly decurrent base; the margins entire or serrate; the petiole base weakly swollen. Infructescence axillary or borne from the trunk. Fruit a berry to 3.5 cm diameter, yellow or orange when mature; the exocarp smooth, thin, fragile; the mesocarp orange. Seeds 1–13 per fruit. Plants glabrous or pubescent, the hairs simple or stellate. Distribution: Nicaragua to Paraguay, to 1200 m elevation.

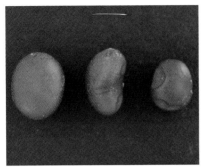

Theophrastaceae - *Clavija* (1.93x; 11.92 × 9.62)

THYMELAEACEAE

Shrubs or herbs; one species a liana. Leaves simple, alternate, the margins entire. Infructescence terminal or axillary. Fruit a drupe. Plants glabrescent or pubescent, the hairs simple.

Schoenobiblus Mart. Shrubs to 3 m tall. Stems with conspicuous lenticels. Leaves thin, nearly transparent when dry; the base strongly decurrent. Calyx persistent. Fruit in groups of 5–8 on a common receptacle, each fruit to 2 cm diameter, pink or reddish when mature. Seeds one per fruit. Distribution: Panama to Peru and Bolivia.

Thymelaeaceae - *Schoenobiblus* (1.67x; 15.13 × 10.82)

TILIACEAE

Trees, sometimes shrubs and herbs. Leaves simple, alternate, with three veins from the base; the margins usually serrate; the petioles pulvinate at the apex. Stipules caducous or persistent. Infructescence axillary or terminal. Fruit an indehiscent capsule or rarely a samara. Seeds arillate or winged. Pubescence of simple or stellate hairs.

Apeiba Aubl. Trees to 30 m tall. Leaves caducous, the base weakly cordate, the margins weakly dentate or serrate. Stipules in a pair at the base of the petiole. Infructescence terminal or opposite the leaf. Fruit a

capsule, to 9 cm diameter, black or brown when mature; appearing dehiscent but opening through an apical, circular pore; the exocarp rather woody with-spine-like projections that are short and indurate or long and soft; the mesocarp white. Seeds many per fruit. Plants pubescent, the hairs simple or stellate. Wood used for lumber. Distribution: Mexico and West Indies to Bolivia, to 1800 m elevation.

Tiliaceae - *Apeiba* (8.99x; 4.55 × 3.39)

Tiliaceae - *Apeiba* (10.95x; 3.5 × 2.63)

Heliocarpus L. Trees to 20 m tall. Leaves entire or sometimes lobed; the petioles variable in size; the base rounded or deeply cordate; the margins finely serrate, with glands on the basal teeth. Stipules present in a pair at the base of the petiole. Infructescence terminal. Fruit indehiscent, to 0.4 cm diameter, surrounded completely by a series of long hairs,

Tiliaceae - *Heliocarpus* (2.62x; 15.93 × 8.95)

brown when mature. Seeds 1–3 per fruit. Plants pubescent, the hairs compound or stellate. Plants typically growing in full sunlight, common in forest clearings and other disturbed areas. Distribution: Mexico to Argentina, to 2000 m elevation.

Lueheopsis Burret. Trees to 25 m tall. Leaves 3-veined from the base, the margins weakly dentate near the lamina apex, the base truncate or weakly cordate. Stipules caducous. Infructescence axillary or terminal. Fruit a 5-valved capsule, dehiscing to only one-third its length from the apex, woody, to 3 cm long, brown when mature, subtended by persistent calyx. Seeds short-winged, many per fruit. Plants pubescent, especially on the abaxial laminar surface that appears somewhat glaucous at first glance, the hairs finely stellate. Plants usually growing in permanently inundated habitats, such as swamp forest. Distribution: Tropical America.

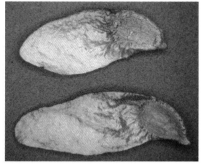

Tiliaceae - *Lueheopsis* (4.89x; 12.25 × 4.24)

Luehea Willd. Trees to 30 m tall with large aerial and buttress roots reaching half the size of the trunk. Leaves 3-veined from the base, the margins serrate especially near the apex, the base asymmetric and weakly cordate. Stipules caducous or persistent. Infructescence axillary or terminal. Fruit a 5-valved capsule, to 3 cm long, opening to only one-third of its length, brown when mature. Seeds winged, many per fruit. Plants pubescent, especially on the adaxial laminar surface; the hairs fine, stellate, and brown; the adaxial leaf surface glabrous and glossy. Distribution: Tropical America.

Tiliaceae - *Luehea* (2.54x; 9.38 × 3.75)

Mollia Mart. Trees to 15 m tall. Leaves 3-veined from the base, the margins serrate. Stipules caducous. Infructescence axillary. Fruit a bivalved capsule, to 1 cm diameter, brown when mature. Seeds short-winged, various per fruit. Plants pubescent. Distribution: Tropical America.

Tiliaceae - *Mollia* (3.07x; 11.79 × 5.28)

ULMACEAE

Trees, sometimes shrubs or lianas. Stems with spines when lianas or with small buttress roots when trees. Leaves 3-veined from the weakly asymmetric base, simple, alternate, or in one genus opposite; the adaxial surface sometimes asperous. Stipules caducous. Infructescence axillary. Fruit a drupe, berry, or rarely a samara. Plants glabrous or pubescent.

Ampelocera Klotzsch. Trees to 30 m tall. Trunk with large buttress roots and exfoliating bark. Leaves coriaceous, with reticulate and somewhat anastomosing venation; the adaxial surface glossy. Stipules small. Infructescence axillary. Fruit a drupe, sessile, to 2.2 cm long, yellow when mature, crowned by persistent stigma; the surface asperous; the mesocarp tan. Seeds one per fruit. Wood used for lumber. Distribution: Mexico to Brazil and Bolivia.

Ulmaceae - *Ampelocera* (2.31x; 19.36 × 12.43)

Celtis L. Trees to 25 m tall, or lianas. Trunk with thin buttress roots that extend various meters through the soil, with clear yellow lenticels of mealy texture in one tree species and with curved, axillary spines in one liana species. Leaves 3-veined from rounded, asymmetric base, the margins entire or serrate, the venation reticulate and weakly anastomosing in some species. Stipules caducous. Infructescence axillary. Fruit a drupe, to 1.7 cm long, yellow to black when mature; the mesocarp yellow, sticky and thick. Seeds one per fruit. Plants variably pubescent throughout, except on adaxial laminar surface. Distribution: North America, Mexico, Central America, and West Indies to Argentina, but more speciose in the temperate zones and in the Old World, to 3300 m elevation.

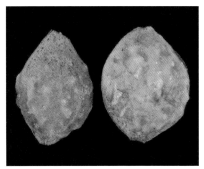

Ulmaceae - *Celtis* (3.87x; 10.11 × 7.92)

Ulmaceae - *Celtis* (5.31x; 11.27 × 5.5)

Ulmaceae - *Celtis* (5.4x; 11.37 × 7.53)

Trema Lour. Trees to 15 m tall. Leaves 3-veined from weakly cordate and sometimes asymmetric base; the apex strongly acuminate; the margins finely serrate or

entire; the adaxial laminar surface often asperous. Stipule caducous. Infructescence axillary. Fruit a drupe, to 0.2 cm diameter, orange or reddish when mature, crowned by persistent stigma. Seeds one per fruit. Plants with light to dense pubescence. Typically found in disturbed areas, such as at the margins of rivers. Bark used as rope and in weaving. Wood used as firewood. Distribution: Southern United States and Mexico to Argentina, to 2400 m elevation, but more speciose in the Old World.

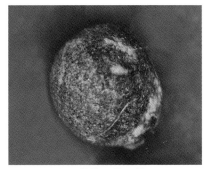

Ulmaceae - *Trema* (28.78x; 1.45 × 1.35)

URTICACEAE

Small trees, shrubs, herbs, or lianas, with one genus of epiphytes. Leaves simple, alternate, 3-veined from the base, sometimes with one small leaf opposite the normal one; the margins usually serrate or sometimes entire or weakly lobed. Stipules small. Infructescence axillary. Fruit an achene or a drupe. Plants often with urticating hairs. The majority of common species of disturbed areas, needing full sunlight.

Urera Gaudich. Shrubs, trees to 6 m tall, or lianas. Leaves 3- to 5-veined from cordate base, the margins entire, widely crenate-dentate, to lobed, the adaxial laminar surface asperous and with white punctations. Stipules axillary. Infructescence axillary or borne from the trunk, the peduncle sometimes dark purple in color. Fruit an achene to 1.2 cm diameter, yellow or orange, transparent or white when mature,

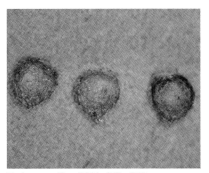

Urticaceae - *Urera* (19.26x; 0.92 × 0.76)

often protected by a persistent, fleshy perianth. Seeds green, nearly exposed, many per fruit. Plants glabrous or with short pubescence throughout, the hairs urticating, sometimes spinose. Distribution: Mexico and the Antilles to Bolivia and Peru, also in Africa and Hawaii, to 2000 m elevation.

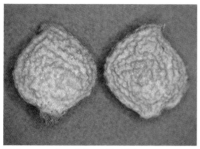

Urticaceae - Urera (8.36x; 4.13 × 3.53)

VERBENACEAE

Herbs, trees, shrubs, and lianas. Stems sometimes square, sometimes with spines. Leaves simple, sessile or petiolate, palmately or trifoliately compound, opposite or verticillate, rarely alternate; some genera with glands at the base of the lamina or on the petiole. Infructescence axillary or terminal. Fruit a berry or drupe, sometimes with the calyx persistent in the form of a cupule, or a schizocarp. Plants sometimes aromatic, especially in the genera of herbs and shrubs.

Aegiphila Jacq. Lianas, shrubs, or rarely trees to 10 m tall. Leaves simple, opposite, and clearly decussate, the venation sometimes anastomosing, the base rounded or weakly cordate. Some species with the leaves grouped at the branch apices. Infructescence terminal or axillary. Fruit a sessile berry, to 2 cm long, orange or red when mature, in compact groups of 5–8, subtended by persistent calyx that sometimes forms a cupule covering up to one-third of the fruit; the mesocarp red. Seeds 1–4 per fruit. Plants glabrous or pubescent, the hairs simple. Distribution: Mexico and the Antilles to Bolivia and Peru, to 3000 m elevation.

Verbenaceae - Aegiphila (6.08x; 8.09 × 3.48)

Verbenaceae - Aegiphila (4.61x; 10.32 × 5.11)

Citharexylum L. Trees to 18 m tall. Stems hollow, lenticellate, square or round. Leaves simple, opposite or 3-verticillate, the margins entire or weakly dentate, the petioles with a pair of conspicuous glands close to the apex. Infructescence axillary and terminal. Fruit a berry, carnose, to 1 cm diameter, red when mature, subtended by persistent calyx generally in the form of a cupule. Seeds two per fruit. Plants common along rivers in the upper Amazon, but also found in the Andes. Distribution: Southern United States to Argentina, to 3600 m elevation.

Verbenaceae - Citharexylum (6.14x; 7.45 × 3.34)

Lantana L. Shrubs to 3 m tall. Stem striate, weakly square or spinose. Leaves opposite or whorled, the margins serrate-crenate; the base de-

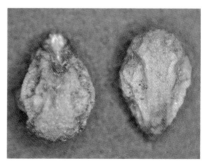

Verbenaceae - Lantana (11.69x; 3.16 × 2.37)

current; the abaxial laminar surface with glossy punctations, the adaxial surface asperous. Infructescence axillary, solitary, 2- to 4- to 6-verticillate. Fruit a berry, to 0.5 cm diameter, dark purple to black when mature, grouped in heads, subtended by brown bracts; the mesocarp white, mealy. Seeds one per fruit. Plants pubescent. Distribution: Southern United States through tropical America, also in Africa.

Petrea L. Usually lianas; also shrubs or trees to 15 m tall. Leaves simple, opposite, rarely verticillate, sometimes asperous. Infructescence axillary or terminal. Fruit a samara, the five elongated calyx lobes persisting to form wings to 3 cm long, blue to violet when immature, brown when mature. Seeds one per fruit. Plants glabrous or pubescent. Distribution: Mexico and West Indies to Peru and Bolivia.

Verbenaceae - *Petrea* (1.82x; 27.62 × 3.02)

Vitex L. Shrubs or trees to 30 m tall. Leaves compound, 3- to 5-foliate, opposite, the petiole generally elongate. Infructescence axillary. Fruit a drupe, to 2 cm long, yellow or dark purple and usually malodorous when mature, subtended by persistent calyx sometimes in the form of a cupule. Seeds one per fruit. Plants glabrous or with pubescence of simple hairs. Some species appreciated for their wood or cultivated as ornamentals. Distribution: Guatemala to Peru and Bolivia, but more speciose in the Old World.

Verbenaceae - *Vitex* (3.01x; 14.61 × 8.73)

Verbenaceae - *Vitex* (3.15x; 16.24 × 7.25)

VIOLACEAE

Shrubs and trees, rarely lianas. Leaves simple, opposite, sometimes alternate, the margins often finely serrate. Stipules usually caducous. Infructescence axillary, terminal, or borne from the stems or trunk. Fruit a capsule or berry.

Gloeospermum Triana & Planchon. Trees to 8 m tall. Leaves alternate, the venation reticulate, the margins serrate. Stipules caducous, leaving a conspicuous circular scar on young stems. Infructescence axillary. Fruit a berry, to 3.2 cm diameter, yellow when mature; the mesocarp yellow, transparent. Seeds 3–27 per fruit. Distribution: Tropical America.

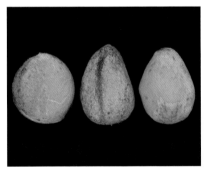

Violaceae - *Gloeospermum* (3.11x; 8.74 × 7.04)

Leonia R.& P. Trees to 20 m tall. Leaves alternate, variable in size, coriaceous, sometimes with reticulate or anastomosing venation. Infructescence axillary, or borne from the stems or trunk. Fruit a berry, to 5 cm diameter, gray-green or yellow when mature; the exocarp subwoody; the mesocarp yellow. Seeds many per fruit. Distribution: Guianas to Brazil and Peru, to 1600 m elevation.

Violaceae - *Leonia* (3.03×; 14.86 × 9.3)

Violaceae - *Rinoreocarpus* (3.28×; 9.97 × 5.48)

Rinorea Aubl. Usually shrubs or trees to 10 m tall. Leaves alternate or opposite, the margins weakly serrate. Infructescence axillary or terminal. Fruit a trivalved capsule, to 2.6 cm long, green or brown when mature, crowned by persistent stigma. Seeds 1–3 per valve. Plants glabrous or pubescent. Distribution: Costa Rica and Panama to Peru and Bolivia, but more speciose in the Old World, to 1800 m elevation.

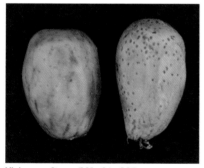

Violaceae - *Rinorea* (5.35×; 8.1 × 5.06)

Violaceae - *Rinoreocarpus* (3.43×; 10.49 × 5.28)

VISCACEAE

Parasitic herbs, shrubs, or epiphytes. Leaves simple and opposite, generally succulent. Infructescence axillary, terminal, or borne from the trunk. Fruit a sessile drupe to 1 cm long, orange when mature. Seeds one per fruit, opening in various segments, usually green with a sticky base. Distribution: Southern United States, Mexico, and West Indies to Argentina.

Phoradendron Nutt. Parasitic shrubs. Leaves coriaceous to succulent. Infructescence axillary. One of the more common and representative genera of the Viscaceae growing in the Amazon. Distribution: Southern United States throughout tropical America.

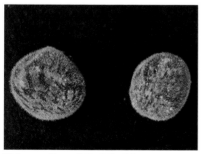

Violaceae - *Rinorea* (3.53×; 7.43 × 7.1)

Rinoreocarpus Ducke. Trees to 10 m tall. Leaves alternate, the margins weakly serrate or crenulate, the venation reticulate and widely anastomosing. Infructescence axillary, terminal, or borne from the stems. Fruit a trivalved capsule, to 3 cm long, yellow when mature, black when dry. Seeds 1–4 per valve. Distribution: Brazil and Peru.

VITACEAE

Lianas. Stems with swollen nodes. Tendrils simple or branched, opposite the leaf. Leaves simple or compound, alternate, the margins entire or lobed. Stipules caducous. Infructescence opposite the leaf. Fruit a berry.

Cissus L. Stems hollow, swollen at each node, square or winged, sometimes striate. Leaves simple or compound, trifoliate or ternate; the base truncate,

entire, or lobed; the petiolules and petioles sometimes weakly winged; the margins finely dentate to
serrate; the venation conspicuously 3-veined from
the base in some species. Infructescence sometimes
with reddish pedicels. Fruit to 2.5 cm diameter, purple or black when mature; the exocarp very thin with
black or white mottling; the mesocarp green or purple, juicy, transparent. Seeds one or two per fruit.
Plants glabrous or pubescent. Distribution: Southern
US, Mexico, and West Indies to Argentina, also in the
Old World.

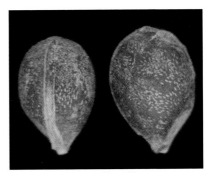

Vitaceae - *Cissus* (6.62x; 6.53 × 4.49)

VOCHYSIACEAE

Trees, rarely shrubs. Leaves simple, opposite or verticillate, the margins entire, the venation usually anastomosing at the margins. Glands or conspicuous
stipules at the petiole base in most genera. Infructescence axillary or terminal. Fruit a loculicidal capsule with winged seeds, or a samara. Plants glabrous
or pubescent, the hairs stellate, T-shaped, or simple.

Erisma Rudge. Trees to 40 m tall. Leaves opposite or
verticillate, coriaceous, sometimes with anastomosing venation or the tertiary venation inconspicuous;
the base rounded or decurrent; some species with
glossy adaxial laminar surface and glaucous or tomentose abaxial surface. Glands or stipules conspicuous at the base of the petiole, sometimes absent. Infructescence axillary or terminal. Fruit a samara, to 9
cm long, the wings four, unequal in size, developed
from the calyx, bluish or brown when mature. Pubescence of stellate hairs. Distribution: South America.

Vitaceae - *Cissus* (5.82x; 4.22 × 3.29)

Vitaceae - *Cissus* (2.56x; 18.47 × 9.59)

Vochysiaceae - *Erisma* (0.79x; 88 × 24.13)

Qualea Aubl. Trees to 30 m tall. Leaves opposite,
sometimes verticillate, coriaceous; the secondary venation perpendicular to the midvein in some species,
very fine and parallel, anastomosing very close to the
margin, very conspicuous on the adaxial surface.
Glands present, raised or depressed, near the junction of the laminar base and petiole apex. Infructescence axillary or terminal. Fruit a trivalved capsule,

Vitaceae - *Cissus* (4x; 9.74 × 7.2)

to 10 cm long, with one thin, central column; brown when mature and subtended by persistent calyx. Seeds winged, various per locule. Plants glabrescent. Wood valued for lumber. Distribution: Tropical and subtropical America, to 1200 m elevation.

Vochysiaceae - *Qualea* (1.86×; 33.52 × 7.9)

Vochysia Aubl. Trees to 35 m tall, rarely shrubs. Leaves opposite or 3- to 4-verticillate; the base strongly decurrent or weakly cordate; the venation anastomosing close to the margin, sometimes inconspicuous. Stipules caducous. Infructescence axillary or terminal. Fruit a trivalved capsule, to 10 cm long, brown when mature. Seeds winged, one to various per locule. Plants glabrous or with golden to reddish pubescence, short and very dense throughout. Wood valued for lumber. Distribution: Tropical America, to 2000 m elevation.

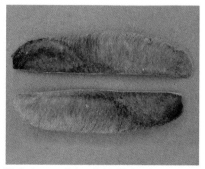

Vochysiaceae - *Vochysia* (1.16×; 53.18 × 12.55)

XYRIDACEAE

Terrestrial or aquatic herbs. Leaves linear, usually cauline and laterally compressed at base, sometimes with purple mottling. Infructescence terminal. Fruit a loculicidal capsule. Seeds many per fruit.

Xyris L. Terrestrial or aquatic herbs, often growing in wetlands; rhizomatous. Leaves linear, usually cauline and laterally compressed at base, sometimes with purple mottling. Infructescence terminal, surrounded by dry papyraceous or indurate

bracts. Fruit a trivalved, loculicidal capsule, about 0.2–0.3 cm long, dry and brown when mature. Seeds many per fruit. Some species cultivated as ornamentals. Distribution: Southern United States and throughout tropical America; also in Australia and Africa.

Xyridaceae - *Xyris* (77.46×; 0.63 × 0.23)

ZINGIBERACEAE

Herbs, usually aromatic when rubbed or crushed; rhizotomous. Leaves simple, alternate, lanceolate; the leaf base sheathing the stem, or leaves petiolate; the venation very fine and parallel. Infructescence terminal or arising from the stem base, usually with colorful, attractive bracts. Fruit a capsule. Seeds often arillate.

Hedychium J. König. Herbs to 1.5 m tall. Leaves distichous, the secondary venation fine and parallel; the petiole base sheathing the stem. Infructescence terminal, the bracts green. Fruit a trivalved capsule, to 3 cm long, orange when mature, subtended by persistent calyx. Seeds many per fruit, with red aril. Plants pubescent, often growing at the margins of streams, near old homesteads, rural commmunities, and roads. Distribution: Naturally from the Himalayas, now widely cultivated and naturalized throughout the tropics of the world.

Renealmia L. f. Herbs to 4 m tall. Rhizomes aromatic. Leaves distichous, the secondary venation

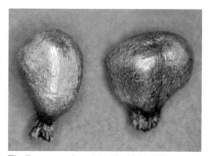

Zingiberaceae - *Renealmia* (4.81×; 5.1 × 5.76)

Zingiberaceae - *Renealmia* (6.19×; 6.07 × 4.89)

very fine and parallel, the petiole base sheathing the stem. Infructescence terminal or arising from the stem base. Fruit a trivalved capsule, to 3.5 cm long, weakly striate, red when immature, black when mature, subtended by conspicuous, persistent, showy, orange calyx. Seeds many per fruit, with orange to red aril. Plants glabrous or pubescent. Distribution: Mexico and the Antilles to Peru and Bolivia, also in Africa.

Glossary and Illustrations of Botanical Terminology

* Botanical terms included in the illustrations.

Abaxial (Plate I-49).* Positioned on the side facing away from the axis, such as the lower leaf surface (compare with ADAXIAL).

Acaulescent. Lacking a stem, or stem so short that it is not apparent, such as in plants with a basal rosette of leaves.

Achene. A small, dry, fruit that does not open at maturity, with a single seed free from the ovary wall, typical of the sunflower family (Asteraceae).

Acuminate. Tapering to a long, sharp point.

Adaxial (Plate I-50).* Positioned on the side facing toward the axis, such as the upper leaf surface (compare with ABAXIAL).

Aerial roots. Roots occurring above the level of ground or water.

Aggregate fruit. A fruit formed by the clustering of separate carpels originating from a single flower, such as blackberries.

Alternate (Plate I-10).* Bearing one leaf per node (compare with OPPOSITE and VERTICILLATE).

Anastomosing venation (Plate I-52).* Leaf veins forming an intertwined network of cross veins connected by a marginal vein.

Anisophilous/anisophylly. A state in which one of two opposite leaves at a node is reduced in size and much smaller than the other leaf.

Anthesis. The time during the flowering period when the flower is open and fully expanded.

Anthocarp. A structure that includes a fruit fused with some portion of the flower, such as the perianth or the receptacle.

Apex (Plate I-45).* The terminating tip of a leaf or the part farthest from the point of attachment.

Apical (Plate I-45).* Positioned at the apex.

Apices. A plural for apex, such as all branch apices on a tree.

Apiculate. Ending in a small, slender point (apicule).

Apocarpous gynoecium (Plate II-19).* The female reproductive organs of a flower (gynoecium) formed from separate carpels.

Aril. A fleshy covering on part or all of a seed, typically edible, thus aiding in seed dispersal by animals.

Arillate. Possessing an aril.

Asperous. Rough to the touch, such as the texture of sand paper.

Asymmetric (Plate I-17).* Without symmetry, not divisible into equal halves, irregular in shape, such as when the base of the leaf does not meet at one point.

Auriculate. Possessing auricles, which are small, ear-like lobes or appendages, found at the base of some leaves.

Axil (Plate I-42).* The point of attachment formed between the axis of a stem and the leaf or fruit arising from it.

Axillary (Plate I-42).* Positioned in, or arising from the axil, such as the position of fruits that arise from the leaf axil where the leaf connects to the stem.

Axis (Plate I-47).* The main, central structure of any plant part.

Basal. Positioned at or arising from the base, such as leaves that arise from the base of the stem.

Berry. A fleshy fruit that develops from a single pistil, with numerous seeds, such as a blueberry.

Bifid (Plate I-2).* Deeply two-lobed from the tip or apex.

Bipinnate/ly (Plate I-22).* Referring to a compound leaf that is twice pinnate with the divisions again pinnately divided.

Biternate (Plate I-23).* A leaf that is doubly ternate with the ternate divisions again ternately divided.

Bract. A reduced leaf or leaf-like structure at the base of a flower or fruit.

Bracteoles. Small bracts borne on a secondary axis, such as a stalk.

Bristles. Short, stiff hairs or hair-like structures.

Buttress roots. Aerial props or supports, often flattened and flaring out from the base of a tree trunk.

Caducifolious. A plant with caducous leaves.

Caducous. Falling off early or prematurely, such as stipules that fall off as leaves emerge, or old leaves that fall slowly over time as new leaves emerge.

Caespitose. Growing in dense clusters, such as leaves clustered at the apex of a branch.

Calyx (Plate II-9).* The outer, leaf-like whorl of a flower; collective term for the sepals of a flower, sometimes attached to the fruit.

Canaliculate. With longitudinal channels or grooves.

Capitula. Small, head-like structures, such as in the clustered flower heads characteristic of the Asteraceae (sunflower) family.

Capsule. A dry fruit opening at maturity, composed of more than one carpel.

Carnose. With a fleshy, meat-like texture.

Carpel. The female reproductive structure in flowering plants; a simple pistil formed from one modified leaf, or that part of a compound pistil formed from one modified leaf. The carpel number of a compound pistil is determined by counting the number of stigmas, styles, locules, and placentae.

Caryopsis. A dry, one-seeded, fruit that does not release its seed, with the seed coat fused to the wall of the fruit (pericarp), as in the fruits of the grass family; also referred to as a grain.

Cauline. Of, on, or pertaining to the stem, such as leaves arising from the stem above the ground.

Ciliate (Plate I-27).* Possessing a fringe of hairs along the leaf margin.

Comose. Possessing a coma or tuft of hairs, such as on the tip of a seed.

Compound (Plates I-18 to I-23).* A leaf that is divided into two or more distinct leaflets with a common stem.

Cordate (Plate I-5).* A heart-shaped leaf with a notch at the base.

Coriaceous. Possessing a tough, leathery texture.

Crenate (Plate I-28).* Leaves that have smooth, rounded teeth along the margin.

Crenulate (Plate I-29).* With very small rounded teeth along the margin.

Cupule. A cup-shaped structure, such as the acorn of an oak tree or the covering on the base of fruits in the Lauraceae (laurel) family.

Curvate. A curved form.

Cystolith. A stone-like mineral mass (concretion), usually of calcium carbonate, which often forms bumps or rounded projections that are visible to the human eye or obvious to the touch, such as in leaves of the Acanthaceae family.

Deciduous. Falling off annually, such as leaves from a tree all falling at the same time; not persistent.

Decurrent (Plate I-15).* A leaf base that extends downward along the stem.

Decussate (Plate I-39).* Leaves arranged along the stem in opposite pairs, with each pair at right angles to the pair above or below.

Dehiscent. Opening at maturity to release the contents, such as fruits that open to release seeds (compare with INDEHISCENT). Related to dehiscence or dehiscing.

Dendritic (Plate I-30).* Hairs or trichomes that have a branching, tree-like form.

Dentate (Plate I-24).* Leaves that are toothed along their margins, with the teeth directed outward rather than forward or upward.

Dimorphism (dimorphic). With two forms, such as two leaf forms on a single plant.

Dioecious. A plant with flowers that are unisexual with the staminate (male) and pistillate (female) flowers borne on different plants (compare with MONOECIOUS).

Discoid. In the form of a disc.

Distichous (Plate I-10 & I-11).* Leaves arranged in two vertical rows on opposite sides of the stem axis; also referred to as two-ranked.

Domatia. Small structures or chambers produced by plants to harbor arthropods such as ants and mites.

Drupe. A fleshy fruit that does not open at maturity, with a stony endocarp surrounding a single seed, such as a peach or cherry.

Drupelet. A small drupe, such as the individual segments or carpels of a blackberry fruit.

Elongate. A seed that is longer than wide.

Endocarp (Plate II-22).* The inner layer a fleshy fruit (compare with EXOCARP, MESOCARP, and PERICARP).

Endosperm (Plate II-24).* The nutritive tissue surrounding the embryo of a seed in flowering plants.

Entire (Plate I-25).* Leaf margins that are continuous and not toothed or divided.

Epipedunculate. Occurring on top of the stalk or peduncle.

Epiphyte/epiphytic. A plant that grows upon another plant for physical support.

Exocarp (Plate II-20).* The outer layer or skin of a fleshy fruit (compare with ENDOCARP, MESOCARP, and PERICARP).

Fenestrate. A tree trunk that possesses window-like pits, openings, or perforations (fenestrae).

Ferrugineous. Reddish-brown, rust-colored.

Follicle. A dry, dehiscent fruit composed of a single carpel and opening along a single side.

Funiculus. The stock or string-like structure that connects the ovule (and seeds) to the placenta of the ovary (fruit).

Glabrescent. Pubescence that is almost hairless (glabrous) or becoming hairless.

Glabrous. Lack of pubescence; hairless, smooth.

Glandarium. A fruit consisting of several to many drupes attached to a common receptacle and originating from a single flower, as in the genus *Ouratea* of the Ochnaceae family treated in this book.

Glandular punctuations. Translucent spots within a leaf, such as the oily, glandular punctuations found along the leaf margin in members of the citrus family (Rutaceae).

Glaucous. With a whitish or bluish waxy coating, such as a glaucous leaf surface or glaucous fruit.

Globose. Spherical, round, globe-shaped.

Gynoecium (Plate II-18).* All of the carpels of a flower, collectively, consisting typically of the ovary (Plate II-15), style (Plate II-16), and stigma (Plate II-17).

Gynophore. An elongated stalk connecting the carpel or pistil to the receptacle in some flowers.

Hemiepiphyte/hemiepiphytic. An epiphyte that is attached to the ground at some point in its life cycle.

Hesperidium. A fleshy, berry-like fruit with a tough, leathery outer skin; refers to fruits in the citrus family (Rutaceae), such as an orange or lemon.

Heterophylly. Possessing different kinds or forms of leaves on the same plant.

Hypanthium. A cup-shaped extension of the floral axis commonly surrounding or enclosing the gynoecium, usually formed from the union of the basal parts of the calyx, corolla, and androecium.

Hypocarp. The swollen stalk in fruits of the genus *Anacardium* (Anacardiaceae), the source of the cashew "nut."

Imbricate. Overlapping like tiles or shingles on a roof.

Indehiscent. Remaining closed at maturity, such as a fruit that does not open to release seeds but rather disperses as one complete, closed unit, such as a fleshy berry eaten and dispersed by a bird (compare with DEHISCENT).

Indurate. Hardened.

Inflorescence. A group of flowers subtended by a leaf.

Infructescence. The fruiting part of a plant originating from the inflorescence and defined as a group of fruits subtended by a leaf.

Imparipinnate (Plate I-20).* The presence of a number of pairs of leaflets plus one extra leaflet at the apex of a compound leaf; a pinnately compound leaf that is odd-pinnate or unequally pinnate.

Involucral. Of or pertaining to an involucre, a whorl of bracts underneath a flower, flower cluster, or the resulting fruiting body.

Lamina (Plate I-44).* The expanded portion of a leaf.

Laminar. Thin, flat, and expanded, as in the blade of a leaf; the laminar surface of a leaf.

Lanceolate (Plate I-1).* Lance-shaped, longer than wide, with the widest point below the middle, as in a lanceolate leaf.

Leaflets (Plate I-46).* Individual leaf-like divisions of a compound leaf.

Legume. A dry fruit or pod that opens at maturity, consisting of one chamber and derived from a single carpel, usually opening along two sides or seams, such as in a green bean or pea pod; a common fruit type found in the bean family (Fabaceae).

Lenticellate. Possessing lenticels, which are slightly raised, somewhat corky, often lens-shaped areas on the surface of stems.

Liana. A woody, climbing vine.

Lianescent. Having the characteristics of a climbing vine.

Linear (Plate I-8).* Long and narrow with more or less parallel margins; resembling a line, such as a blade of grass.

Lobe. A rounded division or segment of an organ, such as in a leaf or a flower petal.

Lobed (Plate I-4).* A leaf or other plant organ that possesses one or more lobes; in some cases leaves can be two-lobed or trilobed.

Locule. The chamber or cavity of an organ, such as the chamber of a fruit containing seeds.

Loculicidal. A fruit that opens through the locules, such as a capsule that breaks at each locule to release seeds.

Loment. A modified legume that is constricted between the seeds.

Malpighiaceous trichomes (Plate I-35).* Plant hairs or trichomes that are T-shaped, found in the Malpighiaceae and other families of flowering plants.

Mealy. Possessing the consistency or texture of flour; powdery, dry, and often crumbly.

Mesocarp (Plate II-21).* The middle layer of the pericarp of a fruit (compare with ENDOCARP and EXOCARP).

Monocarp. A fruit consisting of one carpel, one locule, and one seed; often used to describe the individual fruiting bodies of an apocarpous fruit, such as in the Annonaceae family where numerous separate monocarps are fused to a single receptacle, all originating from a single apocarpous flower.

Monocarpic. Descriptive of a plant that grows to maturity, flowers, and bears fruit only once and then dies, such as the banana (*Musa*) and some species of *Tachigali* in the Fabaceae family.

Monoecious. A plant with unisexual flowers with the staminate (male) and pistillate (female) flowers borne on the same plant.

Monopodial. A branching pattern where the branches arise from a single main axis; a monopodial tree gives rise to branches from only one single trunk.

Mucronate. A leaf apex that is tipped with a short, sharp, abrupt point (mucro).

Multiple fruit. A fruit formed from several to many separate flowers clustered or crowded together on a single axis, such as a pineapple.

Muricate. A surface that is rough with small, sharp projections or points.

Myrmecophilous. A plant or plant organ that has an association or symbiotic relationship with ants, such as the many plants in the Amazon that harbor ants in special myrmecophilous chambers (domatia).

Nectary. A tissue or organ which produces nectar, as in the nectary of a flower.

Ocrea. An opaque or translucent sheath around the stem formed from the stipules, such as in many members of the Polygonaceae family.

Operculum. A small lid that opens or falls off, such as the deciduous cap of a capsule.

Opposite (Plate I-11).* Leaves positioned in pairs across from one another at the same node along a stem; two leaves per node.

Ostiole. A small hole or opening; the opening in the multiple fruit of a fig through which fig wasps enter to pollinate and to breed.

Ovary (Plate II-15).* The portion of the female reproductive organ in a flower which contains one to many ovules; the ovary matures into the fruit and the ovules mature into seeds.

Palmate (Plates I-6 & I-19).* The veins of a simple leaf (Plate I-6) or arrangement of the leaflets of a compound leaf (Plate I-19) that resemble the shape or pattern of the fingers radiating from the palm of a hand.

Paniculate (Plate II-4).* In the form of a panicle, which is a branched, flower cluster maturing from the bottom upwards.

Pappus. The modified calyx of the Asteraceae family, consisting of awns, scales, bristles, or hair-like appendages at the apex of the achene, such as in a dandelion.

Papyraceous. Paper-like in texture.

Paripinnate (Plate I-21).* A compound leaf that is even-pinnate or equally pinnate; typically easy to identify by presence of pairs of leaflets.

Pedicel (Plate II-8).* The stalk of a single flower in an inflorescence.

Peduncle (Plate II-7).* The stalk of a solitary flower of an inflorescence.

Peltate (Plate I-13).* A leaf that is borne on a stalk attached to the center of the lower leaf surface rather than to the base or margin.

Pepo. A fleshy, tough-skinned, indehiscent, many-seeded fruit with three carpels; the fruit of the Cucurbitaceae family, as in a melon or cucumber.

Perfoliate (Plate I-14).* A leaf with the base and margins entirely surrounding the stem, so that the stem appears to pass through the leaf.

Perianth (Plate II-11).* The outer envelope of a flower consisting of the calyx (sepals) (Plate

II-9) and corolla (petals) (Plate II-10) taken collectively.

Pericarp (Plate II-23).* The outer tissue of a fleshy fruit surrounding the seed, consisting of the ENDOCARP (in a drupe), MESOCARP, and EXOCARP.

Persistent. Plant parts such as sepals, petals, bracts, or stipules that remain attached to the fruit at maturity.

Petiolate. A leaf that possesses a petiole.

Petiole (Plate I-43).* The stalk of a simple leaf, connecting it to the stem.

Petiolule (Plate I-48).* The stalk of a leaflet in a compound leaf that connects the leaflet to the axis of the leaf.

Pinna (pinnae, plural**).** One of the primary divisions or leaflets of a pinnately compound leaf.

Pinnate/pinnately (Plates I-21 & I-16).* A compound leaf with leaflets arranged on opposite sides of an elongated axis (Plate I-21), or leaf venation resembling the pattern of a feather with one central vein and numerous secondary veins branching outward (Plate I-16).

Pixis. A fruit that opens at maturity (dehisces) by losing a top lid or cover, opening or exposing a hole or pore from which seeds fall.

Plumose (Plate I-31).* Having hairs or bristles that resemble a feather.

Pneumatophoric. Characteristic of roots of mangrove trees that arise from below water to absorb air from the atmosphere.

Pome. A fleshy, indehiscent fruit consisting of a core (where seeds are located) surrounded by a juicy tissue; often the dry flower parts are visible on the end of the fruit opposite the stalk; an apple.

Pseudobulb. A bulb-like thickening on the stems of many epiphytic orchids.

Puberulent (Plate I-33).* Minutely pubescent with fine, short hairs.

Pubescence. Trichomes or hairs of various forms covering a plant.

Pubescent. Covered with trichomes or hairs.

Pulvinulus (Plate I-38).* Swollen thickening at the base or apex of the petiolule of the leaflet of a compound leaf.

Pulvinus (Plate I-37).* Swollen thickening at the base or apex of a petiole of a simple or compound leaf.

Pulvinulate. Possessing pulvinus or pulvini (plural).

Punctation. Pits or translucent, sunken glands or colored dots found in leaves and other plant organs.

Raceme (Plate II-3).* An unbranched, elongated inflorescence with stalk-like flowers maturing from the bottom upwards.

Rachis (Plate I-47).* The main axis of a structure, such as in a compound leaf or an inflorescence.

Receptacle. The often swollen portion of the stalk that attaches to the flower.

Reflexed. Bent backward or downward.

Reticulate (Plate I-16).* Leaf venation in the form of a network.

Rhizomatous. A plant that possesses rhizomes.

Rhizomes. A horizontal underground stem.

Rugose. Wrinkled.

Ruminate. Roughly wrinkled, with divisions, as in a ruminate endosperm.

Saccate. Nectar glands shaped like a sack or vase.

Sagittate (Plate I-9).* A leaf that is arrowhead-shaped, with the basal lobes pointed downward.

Samara. A dry, indehiscent, winged fruit.

Samaroid. Samara-like.

Saprophytic. A plant living on dead organic matter, lacking chlorophyll; a saprophyte.

Scabrous. Rough to the touch, like sandpaper.

Schizocarp. A dry, indehiscent fruit which splits into separate one-seeded segments at maturity.

Scorpioid. An inflorescence shaped like a scorpion's tail.

Septicidal capsule. A capsular fruit that releases seeds through the septum (septa) and between the locules.

Septum. The partition separating the locules of an ovary or fruit.

Serrate (Plate I-26).* Leaf margins that are saw-like, toothed along the margins with the sharp teeth pointing upward.

Sessile (Plate II-2).* Attached directly without a supporting stalk, such as a leaf without a petiole that is attached directly to the stem.

Simple hairs (Plate I-32).* Undivided hairs with one axis.

Simple leaves (Plates I-10 & I-11).* Leaves with

one blade or lamina not divided into separate leaflets.

Spadix (Plate I-40).* A spike with small flowers crowded on a thickened axis; the characteristic spicate inflorescence of the Araceae family; often covered or enclosed by the spathe.

Spathe (Plate I-41).* A large hood-like bract or pair of bracts attached below and often covering or enclosing an inflorescence, such as in the Araceae family (see SPADIX).

Speciose. Diverse in number of species.

Spicate (Plate II-1).* An inflorescence arranged in a spike.

Spike. An unbranched, elongated inflorescence with sessile or subsessile flowers lacking stalks and maturing from the bottom upwards.

Spikelet (Plate II-5).* A small spike; the characteristic flower cluster of the grass family (Poaceae), consisting of one to many flowers subtended by bracts (glumes).

Spinose. Bearing spines, as in some leaf margins, stems, or trunks.

Sprawling. A plant growth habit characterized by bending or curving downward and creeping along the ground.

Stamen (Plate II-14).* The male reproductive organ of a flower, consisting of an anther (Plate II-13) and filament (Plate II-12).

Stellate (Plate I-34).* Trichomes or hairs that are star-shaped, with many branches radiating from one point at the base of the main axis.

Stigma (Plate II-17).* The portion of the female reproductive organ in a flower which is receptive to pollen; sometimes remaining attached to the mature fruit.

Stipitate. Borne on a stipe or stalk.

Stipel. A small, stipule-like structure at the base of a leaflet of a compound leaf.

Stipule. One of a pair of leaf-like appendages found at the base of the petiole in some leaves, such as in the Rubiaceae family.

Striate. Marked with fine, usually parallel lines or grooves.

Stoloniferous. Bearing stolons; elongate, horizontal stems creeping along the ground and rooting to give rise to a new plant.

Style (Plate II-16).* The portion of the female reproductive organ in a flower which connects the ovary to the stigma; sometimes attached to the mature fruit.

Subconcentric. A wing that is more or less round, surrounding the seed on all sides.

Subconical. More or less cone-shaped.

Subglobose. More or less globose.

Subopposite. Leaves that are more or less opposite but sometimes easily confused with alternate leaves.

Subpentagonal. More or less pentagonal or five-sided in shape.

Subtending. Attached below a given structure, such as bracts subtending a mature fruit.

Subulate (Plate II-6).* Awl-shaped; long, narrow and pointy with margins folding inward.

Subwoody. More or less woody but sometimes easily confused with herbaceous.

Syconium. The multiple fruit of a fig, consisting of a round inflorescence with a hollow, inverted receptacle bearing flowers internally, opening at the top through an ostiole that allows pollinating fig wasps to enter.

T-shaped (Plate I-35).* Hairs or trichomes shaped in the form of a T; also referred to as malpighiaceous hairs.

Tendril (Plate I-36).* A slender, twining organ used to grasp support for climbing vines and lianas.

Terminal (Plate I-51).* The end point or apex of a branch; a flower or fruit positioned at the end of a branch.

Ternate (Plate I-3).* In threes, such as a leaf that is divided into three leaflets.

Tomentose. Covered by a pubescence of short, matted, or tangled, soft wooly hairs.

Trilobate (Plate I-4).* A leaf or other plant organ with three lobes.

Trichome. A hair or hair-like outgrowth of the epidermis of a plant.

Trifoliate (Plate I-18).* A compound leaf that consists of three leaflets.

Trinerved (Plate I-4).* Leaf venation characterized by three veins originating from the base of the lamina.

Trivalved (Plate II-25). Consisting of three valves, such as in a 3-valved capsular fruit.

Truncate (Plate I-7).* A leaf apex or base that is squared at the end as if cut off.

Tuberculate. Bearing tubercles, which are small

swellings or projections that can occur on the fruit, leaves, or stems.

Two-branched (Plate I-35).* Hairs or trichomes that branch into two.

Uncinate. Trichomes or hairs that are hooked at the apex.

Undulate (Plate I-53).* Leaf margins that are wavy but not so deeply as to be lobed.

Urticating. Plants that possess trichomes or hairs that cause stinging irritation when touched, such as in the Urticaceae family.

Utricle. A small, thin-walled, one-seeded, more or less bladder-like inflated fruit.

Valved (Plate II-25).* A fruit that opens by valves, as in capsules and other dehiscent fruits.

Venation. The vein pattern of a leaf.

Verrucose. Covered with wart-like projections or protuberances.

Verticillate (Plate I-12).* Leaves that are arranged three or more per node in a whorl around the stem; also referred to as whorled.

Plate I

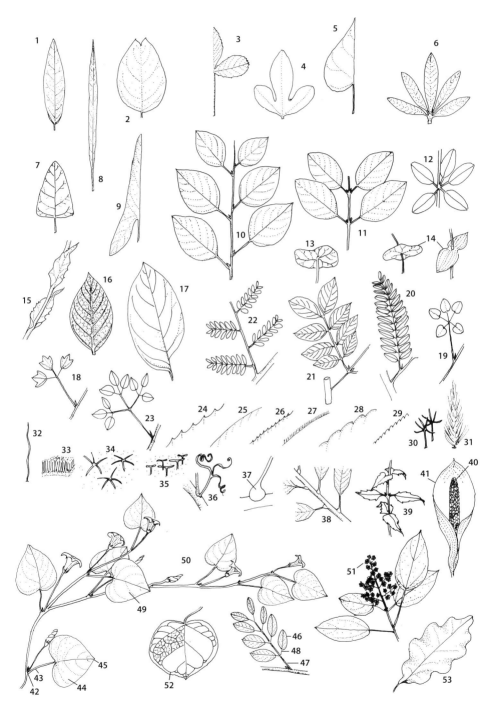

Plate 2

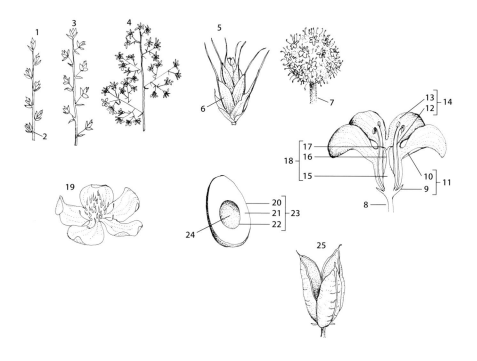

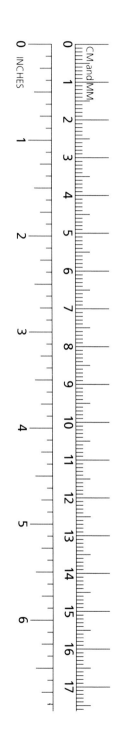

CM and MM

INCHES

6600